Inside OUT

T0330960

Inside OUT: Human Health and the Air-Conditioning Era focuses on the enclosed environment of fully conditioned buildings, revealing a unique ecosystem with broad implications for human life and a rapidly expanding global footprint. Emphasizing the interconnections between buildings and human health, equity, and environmental sustainability, it presents an interdisciplinary, holistic analysis of the social, behavioral, and technological issues of indoor space.

Over the 20th century, advances in mechanical conditioning technologies led to the dispersion and international dominance of the sealed building envelope, which casually and progressively disconnected buildings and their occupants from local climatic, biological, and cultural environments. At the same time, humans were increasingly pushed indoors by less tangible, socially constructed forces that associated climate control with cleanliness, health, social status, and modernization.

In this volume, a multi-disciplinary group of experts on the indoor microbiome from the fields of biology, anthropology, and architecture come together to thoughtfully reflect on the history, properties, and meaning of indoor air quality in buildings, and to discuss the future of human habitation – with a dominant focus on human health in a post-pandemic world. Taking a human-first approach to health and sustainability, the authors weave together a compelling analysis of social and technological drivers of conditioned space with arguments for future interventions in the built environment.

Amid growing awareness of air quality and climate concerns, *Inside OUT* provides a timely discussion of the relationship between building design and human health, of relevance to professional and academic readers from across the spectrum of the building industry, as well as fields including public health and environmental studies.

Elizabeth L. McCormick is an Assistant Professor of Architecture and Building Technology at the University of North Carolina at Charlotte School of Architecture, as well as a PhD candidate at North Carolina State University's College of Design. She is a licensed architect, educator, and researcher whose work explores healthy, climatically sensitive, and contextually appropriate building design strategies that connect occupants to the outdoors while also reducing the dependence on mechanical conditioning technologies.

Inside OUT

Human Health and the Air-Conditioning Era

Edited by
Elizabeth L. McCormick

Routledge
Taylor & Francis Group

NEW YORK AND LONDON

Designed cover image: © Paulo Sousa/Getty Images

First published 2024
by Routledge
605 Third Avenue, New York, NY 10158

and by Routledge
4 Park Square, Milton Park, Abingdon, Oxon, OX14 4RN

Routledge is an imprint of the Taylor & Francis Group, an informa business

ISBN: 978-1-032-50486-5 (hbk)
ISBN: 978-1-032-49589-7 (pbk)
ISBN: 978-1-003-39871-4 (ebk)

DOI: 10.1201/9781003398714

Typeset in Times New Roman
by Apex CoVantage, LLC

Contents

About the Editor

Elizabeth L. McCormick is an Assistant Professor of Architecture and Building Technology at the University of North Carolina at Charlotte School of Architecture, as well as a PhD candidate at North Carolina State University's College of Design. She is a licensed architect, educator, and researcher whose work explores healthy, climatically sensitive, and contextually appropriate building design strategies that connect occupants to the outdoors while also reducing the dependence on mechanical conditioning technologies. McCormick is a LEED and WELL-Accredited Professional, as well as a Certified Passive House Consultant with over 10 years of experience as a practicing architect. She has worked on a variety of project scales from single-family passive houses to LEED-certified commercial office buildings and campuses.

McCormick was the recipient of the 2021 AIAS/ACSA New Faculty Teaching Award and is an active member of numerous professional and academic organizations, including the American Institute of Architects (AIA), National Passive House Alliance (PHAUS), and she is a board member for the Building Technology Educators Society (BTES).

Contributors

Sarah Haines. "The Microscopic World of Building Science"
Haines is an assistant professor in the Department of Civil & Mineral Engineering at the University of Toronto. She has a PhD in environmental science, and her interdisciplinary research integrates microbiology with building science and engineering to analyze the impact the built environment has on human health. Her work aids in understanding indoor exposures from microorganisms and chemicals, providing for a cleaner and sustainable indoor environment.

Marcel Harmon. "The Building/Occupant Organism"
Harmon is an associate principal at BranchPattern, a high-performance building consultancy firm, where he leads the research and development team. He is an applied anthropologist, evolutionist, and engineer. With a PhD in anthropology, Harmon takes a close look at the highly contextual mix of needs, behaviors, and relationships that influence building operations and performance.

Z Smith. "Designing with Metrics for Indoor Air Quality, Comfort, and Health"
Smith is an architect, principal and director of sustainability and building performance at Eskew Dumez Ripple in New Orleans. His built work includes academic, laboratory, and residential buildings. He brings training and experience in physics and engineering to the field of architecture, and is named as inventor on 10 patents and author on over 50 peer-reviewed scientific publications.

Ulysses Sean Vance. "Respiratory Equity"
Vance is an architect and educator working on the implementation of inclusive design and the architectural characteristics of health equity in domestic and institutional settings. He is an associate professor at Temple University in Philadelphia and cofounder of UVXYZI – a design firm dedicated to visualizing the protocols that service inclusive design agendas and communicating the differences in human capacity.

Introduction

Elizabeth L. McCormick

For the first time in history, humans now occupy a geological epoch driven by anthropogenic activity [1]. Culturally, the Anthropocene is defined by the human dominion of nature, a phenomenon that environmental philosopher Freya Mathews calls anthropocentric triumphalism [2]. Rooted in Judeo-Christian traditions found in the stories of Creation, it was believed that humans were shaped in the image of God (Genesis 1:26), who transcends the laws of nature. This belief separated humans from nature to put them in a position of dominance [3], structuring the foundation for Western culture marked by the aggressive domination [4] and commodification [5] of nature. By the late 17th century, philosophical notions of dominion over nature had merged with significant advances in scientific development, which commercialized the practice of human control to form the modern scientific ideology [6]. Many of the tools that we rely on today, including air conditioning, electricity, and automobiles, were developed as systems to dominate nature. Additionally, the transition from human and animal labor as the primary fuel source to coal in the late 18th century furthered the divide between humans and the environment as the manipulation of nature and exploitation of fossil fuels enabled humans to expand their dominion at a planetary scale [3,7]. In the built environment, this phenomenon is evident in the ubiquitous skylines of sealed glass buildings and artificial indoor environments.

> When we try to control nature, we often come to imagine ourselves outside nature . . . as if we were a species alone, disconnected from the rest of life and subject to different rules. **This is a mistake.**
> —Rob Dunn, *A Natural History of the Future*, 2021

Architectural discourse often supports the narrative that humans and nature are inherently dissimilar, a precarious worldview that has led to the stark separation between inside and out in the built environment. The footprint of the interior environment is substantial – physically, psychologically, and ecologically. With global building and construction growth rates of 3–5% per year [8], the ratio of built space to natural landscape is increasingly shifting. For context, the global land

DOI: 10.1201/9781003398714-1

area occupied by interior space is nearly equivalent to that of flooded grasslands and tropical forests on the planet's surface [9]. Once considered an outlier, the built footprint of Manhattan is nearly three times the area of the land that it was built upon; this density is becoming more common in cities around the world.

Not only has the amount of interior space grown, but the ratio of time that humans spend indoors has shifted drastically since the adoption of air conditioning in the 1900s.[i] According to a report from the US EPA, Americans spend upwards of 90% of their lives indoors. This artificial environment, particularly when controlled by mechanical ventilation, is a unique ecosystem with nutrient cycles, biotic life, and energy flows that impact every aspect of the human experience. In fact, the rather dramatic increase in asthma, allergies, and other quantifiable human impacts suggests that it is due to environmental exposures rather than genetic changes [11]. The therapeutic potentials of the built environment have again been thrust into the forefront with the COVID-19 pandemic. Though we continue to regard COVID-19 as unprecedented, it is actually the fifth worldwide influenza outbreak in the past 100 years.[ii] As recently as 2014, the American Society for Heating, Refrigerating, and Air-Conditioning Engineers (ASHRAE) outlined procedures in preparation for a devastating pandemic, for which we were "long overdue" [14].

The built environment has played an integral role in the history of human health. Light and air were commonly promoted as tools to enhance health and hygiene. This notion was the foundation for open air therapy and the sanitorium movement of the 19th century, which led to the development of a new architectural style that used open, well-lit interiors and large areas of glazing to make buildings appear as light and airy as possible, which will be discussed in Chapter 1. Fresh air remained a strategy for sanitation until the invention of antibiotics and chemotherapy treatment in the mid-20th century. Since the 1900s and the widespread adoption of the germ theory of disease, hygiene has become nearly synonymous with sterilization [15]. However, an emergence of research in the last half-century suggests that developed civilizations have become too clean, governed by an antibiotic worldview [16]. This theory, known as the hygiene hypothesis, currently represents one of the most popular explanations for the dramatic increases in childhood illnesses and microbial resistance to antibiotics [17,18]. In designing 'healthy' buildings, the discourse has revolved around the total elimination of microbial communities, which eliminates organisms that are both non-pathogenic and necessary for healthy and robust immune function [19]. This microbial network is an inevitable and essential component of both human and non-human life; however, buildings often neglect this notion, instead opting for sterile, antiseptic interior spaces. Emerging perceptions of the relationship between the human body and microbes are shifting to a more complex and nuanced understanding of their function in maintaining and promoting health. In the same way that healthy bacteria can be used to aid with gut health in the body, there is a growing cohort of researchers who propose a probiotic approach to architecture, "rather than the typical antiseptic, kill all germs method" [20]. The use of beneficial bacteria in buildings has clear benefits; however, it is often met with significant resistance from building occupants and managers due to underlying associations with germs and hygiene.

Competing Imperatives

The obvious solution seems to be to improve ventilation in buildings – enhancing access to outdoor air while maintaining an appropriate level of human control. However, increasing ventilation rates to the levels necessary to create a healthy building would have tremendous impacts on the energy consumption of that building. While many technologies can improve indoor air quality, a technocentric approach will not challenge our relationship with isolated conditioned environments. Similarly, the widespread adoption of mechanical conditioning as the primary thermal strategy and rapid technological advancement as the dominant approach to sustainable development has given way to a ubiquitous placelessness among modern buildings, an overall detachment from nature, and excessive energy consumption, carbon emissions, and expansive ecological degradation. So, how do we proceed? The goal of this book is to promote design changes that purposefully increase biodiversity while also destigmatizing the non-human elements of architecture, leading to more creative solutions that don't rely exclusively on technology.

> When we try to think about reducing the environmental impacts of technology, we tend to focus on the technology itself, but that's not always the most impactful intervention. Often, we have to step back and think through why we're using the service in the first place.
> —Shelie Miller, sustainability scholar [21]

Operating in a cultural context where the concepts of hygiene, purity, and cleanliness need to be reframed, this book supports the belief that humans need to "formulate a new relationship with nature – one based on partnership rather than domination" [6]. However, many of the modern habits associated with the built environment and mechanical conditioning technologies are rooted in psychological associations that are culturally constructed. The human relationship with dirt, germs, and cleanliness, for example, is actually a relatively new construct that stems from social theories of health and hygiene, as will be discussed in Chapter 2. By understanding the dynamics between architecture, mechanical ventilation, and the metrics, this book seeks to provoke alternative models of human habitation that blur the edges between inside and out. By studying the close associations between sociocultural constructs and technological development, *InsideOUT* serves to contextualize modern strategies of indoor conditioned space within the broader view of global sustainability and human health. The indoor condition exists in an interdependent web of circumstances that stretch well beyond its immediate disciplinary boundaries – the human relationship with indoor space is complex and cannot simply be studied through a single method or disciplinary lens. In response, this book brings together ecologists, anthropologists, and architects to discuss the state of interior space in the air-conditioning era, with the hope that architecture can again be used as a tool that makes humans well, instead of one that just keeps us from getting sick.

Notes

i   This transition occurred concurrently with significant economic shifts from manufacturing to knowledge-based sectors, which increasingly took place in indoor environments [10].

ii In 1918, there was H1N1 (Spanish flu, 50 million+ people died); H2N2 (Asian Flu, 1.1 million people died) in 1957; H3N2 (Hong Kong Flu, 1 million people died) in 1968; and H1N1 again in 2009 (Swine Flu, 284,000 deaths). Strategies used to combat the Spanish flu (1918) included wearing cloth masks, hand washing, social distancing, quarantine and self-isolation, contact tracing, and banning of public gatherings [12]. In response to these types of outbreaks, mechanical ventilation strategies, including dilution and exhaust ventilation, ultraviolet germicidal irradiation, and central system filtration were designed to prevent the spread of airborne diseases through HVAC systems in indoor spaces [13].

References

[1] Crutzen, Paul J. "Geology of Mankind." *Nature* 415, no. 6867 (2002): 23. https://doi.org/10.1038/415023a.

[2] Mathews, Freya. "Towards a Deeper Philosophy of Biomimicry." *Organization & Environment* 24, no. 4 (December 1, 2011): 364–87. https://doi.org/10.1177/1086026611425689.

[3] Geisinger, Alex. "Sustainable Development and the Domination of Nature: Spreading the Seed of the Western Ideology of Nature." *Boston College Environmental Affairs Law Review* 27, no. 1 (1999/2000): 43–74.

[4] Bauckham, Richard. "Modern Domination of Nature." In *Environmental Stewardship*, 32–50. London: A&C Black, 2006.

[5] Huber, Matthew T. "Energizing Historical Materialism: Fossil Fuels, Space and the Capitalist Mode of Production." *Geoforum, Themed Issue: Postcoloniality, Responsibility and Care* 40, no. 1 (January 1, 2009): 105–15. https://doi.org/10.1016/j.geoforum.2008.08.004.

[6] Merchant, Carolyn. *Reinventing Eden: The Fate of Nature in Western Culture.* London: Taylor & Francis Group, 2013.

[7] Steffen, Will, Jacques Grinevald, Paul Crutzen, and John McNeill. "The Anthropocene: Conceptual and Historical Perspectives." *Philosophical Transactions of the Royal Society A: Mathematical, Physical and Engineering Sciences* 369, no. 1938 (March 13, 2011): 842–67. https://doi.org/10.1098/rsta.2010.0327.

[8] United Nations Environment Programme. *2020 Global Status Report for Buildings and Construction: Towards a Zero-Emissions, Efficient and Resilient Buildings and Construction Sector.* Nairobi: UNEP, 2020.

[9] Martin, Laura J., Rachel I. Adams, Ashley Bateman, Holly M. Bik, John Hawks, Sarah M. Hird, David Hughes, et al. "Evolution of the Indoor Biome." *Trends in Ecology & Evolution* 30, no. 4 (April 1, 2015): 223–32. https://doi.org/10.1016/j.tree.2015.02.001.

[10] gl Horr, Yousef, Mohammed Arif, Amit Kaushik, Ahmed Mazroei, Martha Katafygiotou, and Esam Elsarrag. "Occupant Productivity and Office Indoor Environment Quality: A Review of the Literature." *Building and Environment* 105 (August 15, 2016): 369–89. https://doi.org/10.1016/j.buildenv.2016.06.001.

[11] Sundell, J. "On the History of Indoor Air Quality and Health." *Indoor Air* 14, no. s7 (August 2004): 51–8. https://doi.org/10.1111/j.1600-0668.2004.00273.x.

[12] Gilbert, Heather A. "Florence Nightingale's Environmental Theory and Its Influence on Contemporary Infection Control." *Collegian, Special Issue on Nursing and Health Care History* 27, no. 6 (December 1, 2020): 626–33. https://doi.org/10.1016/j.colegn.2020.09.006.

[13] ASHRAE. *ASHRAE Position Document on Airborne Infectious Diseases*. Atlanta, GA: ASHRAE, January 19, 2014.

[14] Schoen, Lawrence J. "What ASHRAE Says About Infectious Disease." *ASHRAE Journal* (November 2014): 3.

[15] Vandegrift, Roo, Ashley C. Bateman, Kyla N. Siemens, May Nguyen, Hannah E. Wilson, Jessica L. Green, Kevin G. Van Den Wymelenberg, and Roxana J. Hickey. "Cleanliness in Context: Reconciling Hygiene with a Modern Microbial Perspective." *Microbiome* 5, no. 1 (July 14, 2017): 76. https://doi.org/10.1186/s40168-017-0294-2.

[16] Greenhough, Beth, Andrew Dwyer, Richard Grenyer, Timothy Hodgetts, Carmen McLeod, and Jamie Lorimer. "Unsettling Antibiosis: How Might Interdisciplinary Researchers Generate a Feeling for the Microbiome and to What Effect?" *Palgrave Communications* 4, no. 1 (December 11, 2018): 1–12. https://doi.org/10.1057/s41599-018-0196-3.

[17] Blaser, Martin J. *Missing Microbes: How the Overuse of Antibiotics Is Fueling Our Modern Plagues*. First Picador edition. New York: Picador, 2015.

[18] Scott, Elizabeth A., Elizabeth Bruning, Raymond W. Nims, Joseph R. Rubino, and Mohammad Khalid Ijaz. "A 21st Century View of Infection Control in Everyday Settings: Moving from the Germ Theory of Disease to the Microbial Theory of Health." *American Journal of Infection Control* 48, no. 11 (November 1, 2020): 1387–92. https://doi.org/10.1016/j.ajic.2020.05.012.

[19] Beckett, Richard. "Probiotic Design." *The Journal of Architecture* 26, no. 1 (January 2, 2021): 6–31. https://doi.org/10.1080/13602365.2021.1880822.

[20] Gray, Audrey. "After 2020, Designing for Indoor Air Quality Will Never Be the Same." *Metropolis*, December 21, 2020. https://metropolismag.com/viewpoints/designing-for-indoor-air-quality/.

[21] Bolahke, Suaget. "Rethinking Air Conditioning amid Climate Change." *Ars Technica*, May 28, 2022. https://arstechnica.com/science/2022/05/rethinking-air-conditioning-amid-climate-change/.

1 Air Quality and Human Health

Elizabeth L. McCormick

As deaths from major infectious diseases like AIDS, tuberculosis, and malaria decline, other preventable killers become more visible. The harm caused by air pollution and exposure to hazardous environmental chemicals is **the new epidemic,** demanding urgent attention in the era of sustainable development. **Public health cannot tackle a problem of this magnitude using conventional tools like vaccines and medicines . . .** we also need full engagement from the energy, transport, and finance sectors.

—Dr. Margaret Chan, Director General of
the World Health Organization [1]

When one thinks of air pollution, they are likely to picture billowing clouds from smokestacks, car exhaust, and maybe even a smoggy sunset. Though the greatest dangers are largely attributed to ambient (outdoor) air, household air pollution is the second-greatest environmental threat globally, with an estimated three billion people exposed to poor indoor air from the use of solid fuels like coal or biomass for cooking, heating, and lighting [2]. Although this number has decreased significantly over the past thirty years, exposure to low-quality air remains a significant health concern in developing countries, with major effects on respiratory and cardiovascular health [3]. In heavily industrialized countries with reliable energy infrastructure such as in the United States, however, there is limited concern for the use of solid fuels indoors, and ambient air pollution has gone down significantly since the introduction of the Clean Air Act, the federal law that regulates air pollution in the United States, in 1970. However, indoor air poses an entirely unique threat in industrialized countries, particularly as construction materials are using a greater level of synthetic building materials which can emit contaminants indoors, some of which are still unknown.

According to the EPA, the quality of indoor air can be two to five times worse than outdoor environments, especially with regard to airborne chemicals [4].

DOI: 10.1201/9781003398714-2

It's possible, and even likely, that the air in the middle of a busy intersection might actually be cleaner than the air in your living room right now [5]. However, most Americans still perceive the risks of outdoor air pollution as being substantially higher than the threat of indoor air [6], even though dangers posed by long-term exposure to unhealthy indoor air have become more apparent in recent years, particularly for people who suffer from allergies and asthma, as well as children and the elderly. Allergies are increasing across the globe, but more so in the developed world due to chemical exposures in low-quality indoor air. A growing body of evidence shows that the rather dramatic increase in quantifiable human impacts implies that it is due to environmental exposures, *not* genetic changes [7]. Additionally, a recent study published by *BMJ*, a leading medical journal, demonstrates that exposure to air pollution, particularly particulate matter, is associated with an increased risk of clinical dementia and cognitive decline [8], even with exposure levels below the current EPA standards.[i] According to a report from the US EPA, Americans spend 90% of their time indoors, with higher levels for children and the elderly, the most vulnerable groups to low-quality air. However, the effects of long-term exposure to indoor contaminants may not emerge for many years, and if they do, their source may not be immediately obvious.

Soon after the superinsulation and energy efficiency movement of the 1970s, the World Health Organization (WHO) identified the health problems associated with indoor spaces as Sick Building Syndrome (SBS) in 1983. SBS is described as a set of physical, chemical, and psychological factors affecting human health and comfort, including asthma, allergies, watery eyes, cough, headache, and fatigue, while more severe ailments can include cardiovascular issues, cancers, and reproductive problems [10,11]. Studies show that upper respiratory symptoms and fatigue are more common in sealed buildings with HVAC compared to naturally ventilated buildings [12]. Aside from individual health implications, a report from the EPA estimated that poor indoor air costs the US *billions* of dollars in lost productivity and medical expenses per year [13]. Other studies indicate that the annual costs associated with productivity losses from SBS in the commercial workplace are estimated to reach up to $70 billion per year in the United States alone [14]. On the other hand, annual savings (productivity gains) from better indoor environments can range from $10–20 billion from reduced SBS symptoms and up to $125 billion from direct improvements to worker productivity unrelated to health [15].

Homes can be built to even stricter standards than commercial buildings and with an entirely different set of indoor contaminants. Studies have found that homes with gas stoves, for example, have indoor concentrations of NO_2 that are 2–3 times higher than homes with electric stoves [16]. According to a report from Lawrence Berkeley National Laboratory, cooking with gas stoves exposes millions of Americans to pollutant levels in excess of quality standards for these pollutants (if such outdoor standards were applied indoors), which is particularly impactful to vulnerable populations like children or the elderly. A meta-analysis of over forty studies published between 1977 and 2013 found that children living in homes with gas stoves had a 42% increased risk of having childhood asthma and a 24% increased risk of lifetime asthma [17]. Additionally, a 2017 report from the Massachusetts Department of Public Health affirmed that gas stoves were the most commonly reported trigger of pediatric asthma symptoms, even more than carpets, rugs, or pet dander, and it is projected that nearly 13% of childhood asthma cases[ii] could have been prevented if gas stoves were not used [18]. Despite these findings, there is not a full national accounting of asthma triggers, so it is still difficult to compare against the health threats of other indoor pollutants. Despite mounting data about the invisible enemy lurking inside our buildings, indoor air pollution remains largely unregulated in the United States.

Ventilation in Buildings

Energy use intensity (EUI) is the most commonly used indicator of energy use in buildings. It is helpful in comparing buildings of different sizes, since it is normalized by building area (kBTU/sf/yr or kWh/m^2/yr). However, there is a growing concern over the blanket use of EUI as a singular gauge of building performance, particularly in regard to human health and embodied carbon, though the latter falls outside the scope of this project. The outdoor air quality index (AQI) is calculated based on the five major air pollutants regulated by the United States' Clean Air Act: ground-level ozone (O3), particle pollution (particulate matter, $PM_{2.5}$ and PM_{10}), carbon monoxide (CO), sulfur dioxide (SO_2), and nitrogen dioxide (NO_2). The pollutants most studied indoors include ozone, carbon monoxide, particulate matter, as well as volatile organic compounds (VOC or TVOC) and formaldehyde (HCHO). Additionally, carbon dioxide (CO_2) can be a good indicator of the general freshness of the air, though it is not largely toxic in itself. Other common concerns are respiratory disease, radon, chemical buildup, secondhand smoke, legionella bacteria, as well as more benign contaminants including mold, dust, pet dander, and insect allergens.

> **Ventilation comes next to godliness,** and must necessarily precede the manifestation of the latter, so I hold that in the next decade one of the great human benefactors will be the man or the body of men who make it possible for their fellowmen to be more cleanly in their surroundings.
> —Edward P. Bates, presidential address at the first meeting of the American Society of Heating & Ventilating Engineers (ASHVE, now ASHRAE) [19]

Prior to the 1930s, ventilation standards were largely governed by the control of smells as body odors from humans were considered the primary source of indoor pollution [7,20]. Buildings were initially designed to 'breathe,' though this was often uncontrolled due to the poor construction of buildings and the unpredictable nature of outdoor air. The oil crisis of the 1970s heightened environmental concerns around the use of fossil fuels for heating and cooling, particularly from foreign sources, prompting a new emphasis on the energy performance of buildings and improved construction standards. Constructed to higher, more airtight standards, this new generation of energy-efficient buildings often limited the use of fresh air to reduce the environmental load associated with mechanical conditioning. Despite energy savings, airtight, mechanically ventilated buildings were often plagued with indoor contaminants impacting human health and comfort from a lack of fresh air exchange. There was limited knowledge of non-industrial indoor air conditions until problems with radon and formaldehyde emerged in the late 1960s [7].

Today, ASHRAE Standard 62, *Ventilation for Acceptable Indoor Air Quality*, is the predominant mechanism used in the United States to regulate ventilation rates in buildings. However, like all ASHRAE standards, it was not considered regulatory until adopted by building codes, and it wasn't until 1989 that it was converted into code-intended language with specific, measurable, and enforceable requirements, despite its initial publication nearly two decades prior [21]. Though the standard was first published in 1973 to preserve "occupants' health, safety and well-being" [22], it wasn't until 1999 that ASHRAE membership voted that their standards should "strive to provide health, comfort and/or occupant acceptability", addressing *human health* in addition to *comfort . . .* where appropriate. As a result, the board of directors approved the inclusion of the disclaimer that all indoor air quality and ventilation standards "shall not make any claim or guarantees that compliance will provide health, comfort or occupant acceptability, but shall strive for those objectives" [21]. Ventilation standards developed from an organization of well-intentioned engineers "steeped in 19th century utopianism" [23] who sought to solve the social problems of the world with technology. It wasn't until May of 2023, as this chapter was being written, that ASHRAE released its first health-based ventilation standard (for review) in response to the COVID-19 pandemic, Standard 241P, *Control of Infectious Aerosols*. The standard provides requirements for ventilation, filtration, and air-cleaning systems to reduce exposure to infectious aerosols.

Current international guidelines, including ASHRAE, require a minimum of 15 CFM per person of fresh air ventilation. However, recent studies, including Joseph Allen's 2015/16 Cognitive Function (COGfx) study, call for 'enhanced' ventilation of at least 40 CFM per person to maintain minimum standards of indoor environmental health. Allen and his team found that cognitive function scores of the professional-grade employees under evaluation were 101% higher in the Green+ building (40 CFM/person, low TVOC) and 61% higher in the simulated green building (20 CFM/person, low VOC), compared with the base (conventional) condition (20 CFM/person, high VOC) [24]. Follow-up studies found that providing enhanced ventilation at a rate of 40 CFM/person would cost less than $40/person/year but improve employee performance by 8%, or $6,500/year [25].

Though ambient air pollution has been federally regulated in the United States for over fifty years, there are still no federal laws to protect indoor air and only two states, California and New Jersey, currently regulate indoor air quality. The Occupational Safety and Health Administration (OSHA), the organization that promotes healthy and safe working conditions in the United States, does not set standards for indoor contaminants; however, it does regulate ventilation rates like ASHRAE. LEED (introduced in 1998) and WELL (introduced in 2013), however, are two international green building organizations that are trying to address the concerns associated with indoor air quality, though participation is still discretionary. During the COVID-19 pandemic, ventilation with outdoor air was the most important strategy to mitigate the risk of transmission of the COVID-19 virus and suggested increasing outdoor air supply to 100% wherever possible [26], a strategy that was agreed upon by the three most prominent international HVAC-related institutions, ASHRAE (US), REHVA (Europe), and SHASE (Japan). Each of the guidelines issued by these institutions emphasized the importance of fresh air; however, the specific ventilation rate for eliminating the transmission of airborne particles has not been concretely defined [27]. Most agencies agreed that it is impossible to fully eliminate the transmission of the COVID-19 virus indoors, even with strict controls of the HVAC system. Instead, infection risk is minimized through proper measures and 'bundle control strategies', including social distancing and personal hygiene.

Therapeutic History of Fresh Air

Though the idea of mechanical ventilation is a relatively *new* concept, the idea that polluted air may be detrimental to human health has been around since the time of the Greeks and Romans. As soon as fire was brought into the home, humans realized the need to vent smoke out and allow air in to keep the fire burning. In fact, the control of combustion provided the first motivation for the ventilation of indoor space [28] though it wasn't studied scientifically until the 18th century [29]. Since then, there has been a relationship between human health and environmental conditions, particularly fresh air and daylight. Greek physician Hippocrates, who is known for developing the Hippocratic Oath (c. 400 BC) to "do no harm", believed that to better understand diseases, medical professionals must consider a patient's physical environment, as well as the cultural context of its inhabitants [30,31]. In fact, it was the ancient Greeks who emphasized holistic perceptions of space by understanding the physical as well as the social context [32]. In more recent history, the relationship between the built environment and medical treatment has been well documented by scholars such as Beatriz Colomina, Margaret Campbell, and others. This relationship is most evident in three areas of history: pediatrics, healthcare, and education, which will be discussed in the following sections.

Pediatrics

In historical parenting literature, fresh air was thought to be "of almost as much importance to the baby as food" [33,34]. In the late 19th century, Dr. Luther Emett

Holt, an American pediatrician and professor of childhood disease, wrote a popular parenting book entitled, *Writing the Care and Feeding of Children: A Catechism for the Use of Mothers and Children's Nurses*. In it, he recommended the process of *airing out children* to "renew and purify the blood." Excerpts from Holt's prominent text include:

Is there not great danger of a young baby's taking cold when aired in this manner?

Not if the period is at first short and the baby accustomed to it gradually. Instead of rendering the child liable to take cold, it is the best means of preventing colds.

Of what advantage to the child is going out?

Fresh air is required to renew and purify the blood, and this is just as necessary for health and growth as proper food.

What are the effects produced in infants by fresh air?

The appetite is improved, the digestion is better, the cheeks become red, and all signs of health are seen [35].

Even government publications from the US Children's Bureau promoted the benefits of fresh air and sunshine for babies and children in the 1930s [36]. By providing regular access to fresh air, infants could theoretically build their immune systems to ward off common colds. In 1922, the baby cage was invented as a solution for people living in cramped apartments who still wanted to *air out* their child [37,38], as shown in Figures 1.1 and 1.2. This device, also known as the 'window crib' or 'balcony cot', was a small outdoor sleeping compartment attached to apartment windows so that urban babies could have access to the healing effects of outdoor air [39].

Healthcare

British nurse and social reformer Florence Nightingale's (1820–1910) practice of environmental theory or environmental hygiene, more commonly known today as *infection control*, acknowledged that an appropriate environment was integral to the recovery of ailing patients. In addition to building the foundation for contemporary nursing practices, Nightingale was responsible for significant evidence-based healthcare reforms that had notable implications on the spatial influence on human health. For context, it wasn't until the 1840s that the simple act of washing hands between patients was known to dramatically reduce the infection rates in hospitals [40]. At the start of her career in the 1850s, little was

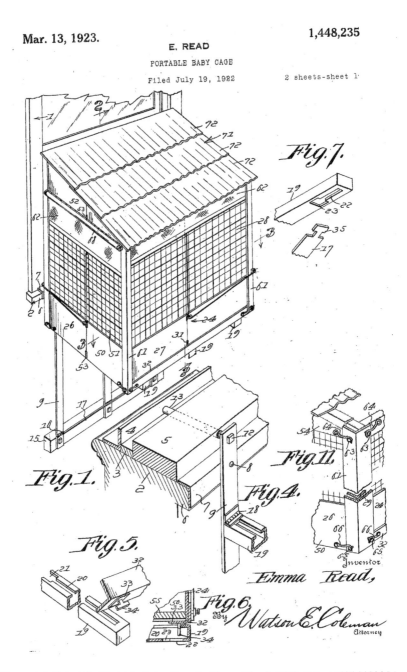

Mar. 13, 1923.

1,448,235

E. READ

PORTABLE BABY CAGE

Filed July 19, 1922

2 sheets-sheet 1

Figure 1.1 Patent for Portable Baby Cage by Emma Read, 1923. Patent US1448235A.

Figure 1.2 Baby cage in London, 1934. Fox Photos. Getty Images, Hulton Archive.

known about the difference between bacterial or viral contagion; however, it was becoming increasingly clear that diseases could be spread through direct or indirect contact, which led her to connect the importance between the spread of disease and the physical environment (pure air, pure water, efficient drainage, cleanliness, and light) [40]. After returning to Britain from the medical barracks that she experienced during the Crimean War, Nightingale was a forceful advocate for pavilion-style hospital designs that would provide fresh air, natural light, and appropriate physical distances between patients, to not inhibit nature's innate recovering processes [41]. Temporary facilities implemented around the world that dealt with the dramatic influx of infectious patients during the COVID-19 pandemic closely resembled Nightingale's open wards of the 19th century.

Tuberculosis (TB) is the airborne infectious disease that plagued urban populations before antibiotics while simultaneously revolutionizing the human relationship between fresh air and human health. TB killed more people between 1914 and 1918 than the first World War [42]. The bacteria causing TB had been around for millions of years before it was officially 'discovered' in 1882 while killing one out of every seven people in Europe and the United States [43]. It wasn't until

1865 that Tuberculosis was understood as an infectious disease, spread through close contact between humans, and not until 1882 was it discovered that it came from the tubercle bacillus. This discovery by Robert Koch is regarded as one of the most significant medical-scientific achievements in human history to date, for which he was awarded the Nobel Prize in Physiology or Medicine in 1905 [44]. Though Koch's discovery had little immediate influence on the reduction of tuberculosis morbidity rates, it provoked a health movement that dramatically shifted the way that medical patients receive care and treatment. Architecturally, the discovery that disease is spread through human transmission led to a movement that was both a therapeutic and social phenomenon [45]. Architectural historian Julie Collins wrote that the hospital became "an instrument of the cure itself", a "form of medical technology" [31].

There was a strong belief in the healing qualities of fresh air and the regenerative power of the sun, a notion that was influential in modernist hygienic ideologies [45]. The phrase *Licht und Luft* (light and air) was commonly used amongst late-nineteenth-century hygiene reformers to promote health and hygiene. This resulted in a new style of architecture that used open, well-lit interiors and large areas of glazing to make buildings appear as light and airy as possible, as shown by the sun parlor in Figure 1.3. In 1840, British physician and pulmonary specialist George Bodington published a study on the treatment and care of pulmonary diseases, where he noted that people who worked outdoors, such as farmers and shepherds, appeared *immune* to Tuberculosis, which was more likely to infect people who spent their time indoors, particularly in urban settings. He reasoned that patients should learn from the lifestyles of outdoor laborers and spend more time breathing fresh outdoor air [46]. This notion was the foundation for *open air therapy* (*open air* method) and the sanitorium movement of the 19th century.

People who had contracted tuberculosis, referred to as *consumptives*, were initially cared for at home by family members. The sanitorium emerged in Europe as a standard treatment facility for long-term illnesses, promoted by both architects and medical professionals "not only as therapeutic instruments but also as instruments of prevention" or "curing machines" [31]. The first sanatorium for the treatment of TB was opened in 1863 by Hermann Brehmen in Silesia (now Poland), where patients were exposed to outdoor air, sunlight, and good nutrition [47]. The first American sanitorium opened 21 years later in 1884 at Saranac Lake in New York, known as the Adirondack Cottage Sanitorium. In addition to the tranquil, wooded setting, deep verandas and covered balconies were critical elements of sanitorium architecture, exposing patients to fresh air while keeping them protected from the wind and rain.

Education

In a similar response to the tuberculosis epidemic, the open-air school movement emerged as a hybrid between sanitorium and schoolhouse, designed to limit exposure to tuberculosis in children who were deemed to be 'pre-tubercular' [48].

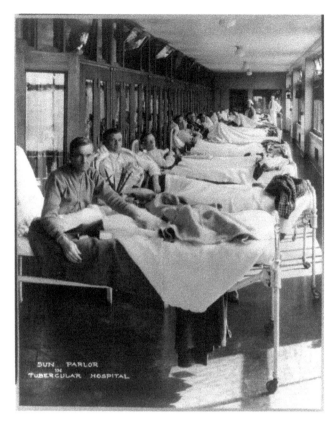

Figure 1.3 Sun Parlor in Tubercular Hospital, ca. 1910–1920. Archived in the Library of Congress Web Archives at www.loc.gov.

Founded by pediatrician Bernhard Bendix, the first open-air school, *Waldschule für kränkliche Kinder* (forest school for sickly children), shown in Figure 1.4, opened in Germany in 1904. It was described as *the school of the future* by Parisian architect Augustin Rey, who believed that the application of contemporary hygienic principles to the nation's school buildings would benefit the children's health, ensuring healthy, vigorous generations of children instilled with "the joy of living, the strength to work and, later, to fight" [49,50]. Interest in the open-air school movement spread quickly to the United States, where two Rhode Island doctors opened the first open-air school in Providence in 1908 for students who had been exposed to tuberculosis. Pediatricians Mary Packard and Ellen Stone converted an empty brick building to an open-air school, with high ceilings and ceiling-height windows on every side, which were kept open. Children kept warm with blankets or "Eskimo sitting bags" with heated soapstones placed at their feet [51]. By 1910, there were 65 open-air schools across the country.

Figure 1.4 First open-air school, The Waldschule für Kränkliche Kinder (forest school for sickly children) in Charlottenburg, Germany, 1904.

Figures 1.5 and 1.6 show two open-air schools in New York City. Building on this success, Dutch architect Jan Duiker designed an open-air school in center of Amsterdam in 1927 to offer the benefits of fresh air to healthy schoolchildren, whether or not they had been infected by tuberculosis. Like sanitoriums, open-air schools lost popularity with the invention of antibiotics in the 1940s, and the last fresh-air school closed in Providence in 1957 [51].

COVID-19 Pandemic

The concept of the fresh-air cure seemed to return (temporarily) with the COVID-19 pandemic. New York City, which had over one million children in public education in 2020, approved roughly 1,100 proposals for public schools to hold classes outdoors during the fall of 2020. Essex Street Academy in New York City, for example, used their roof, which was already designated as a schoolyard, to hold classes [52]. The National COVID-19 Outdoor Learning Initiative was formed in the Spring of 2020 "to once again encourage schools and districts to look beyond their classroom walls". Formed by leaders from environmental education organizations, the initiative created frameworks, strategies, and guidance for school districts to creatively use outdoor spaces for education across the United States. In Maine, the Portland Public School District was able to accommodate more than 5,000 students in outdoor classrooms throughout the year,

Figure 1.5 An Open Air Class on Day Camp Rutherford, New York across river, 1911. Archived in the Library of Congress Web Archives at www.loc.gov.

Figure 1.6 Rest hour for a fresh air class, Public School #51, Manhattan, New York, 1911. Archived in the Library of Congress Web Archives at www.loc.gov.

Figure 1.7 Outdoor Classroom in the fall of 2020 during the COVID-19 Pandemic. Dominican University, River Forest, Illinois.

even in winter with the appropriate 'infrastructure', which included warm clothing, hot drinks, and regular movement [53]. Many of the teachers involved with outdoor teaching during the pandemic believed that these learning environments "brought some much-needed normalcy to pandemic school years" [54]. Now that the risks associated with COVID-19 have dissipated and education has largely returned to 'normal', the group is advocating for outdoor learning as a long-term solution to some of the other systemic inequalities in academic settings.

Navigating an Antibiotic Worldview: The *Probiotic Turn* [55]

Sanitorium treatment required patients to leave their homes for long periods of time. This strategy was cumbersome and was quickly replaced by the convenience of antibiotics and chemotherapy in the 1940s and 50s, which were far less expensive and disruptive than a stay in a sanatorium. Nobel Laureate Paul Ehrlich described the advent of effective chemotherapy as a *magic bullet* in that it could selectively target disease-causing microorganisms without killing their human host [44]. Consequentially, the belief in the curative properties of outside air and environmental space diminished, breaking the direct association between the built environment and human health and dramatically disrupting the relationship between humans and the environment [45,56,57]. Instead, humans now attempt to control disease through *medicine* and a "pill for every ill" [58], addressing health through technology (treatment) instead of salutogenic (preventative) measures.

The dramatic shift to an antibiotic lifestyle and the techno-centric approach as the dominant model for healthcare in industrialized countries has translated directly into the built environment, with mechanical ventilation used to provide sterile and predictable interior environments. In true antibiotic fashion, the discourse surrounding the modern approach to 'healthy' buildings has revolved around the total elimination of microbial communities, which eliminates organisms that are both non-pathogenic and necessary for healthy and robust immune function [59]. This interior microbiome is an inevitable and essential component of human health, which will be discussed in Chapter 3 (Haines). However, buildings often neglect this notion and opt for sterile, antiseptic interior spaces. According to architectural scholar Carolina Ramirez-Figueroa, architectural thought clings to the antibiotic turn, lagging well behind medical discourse [55].

Emerging research in the last half-century suggests that developed civilizations have become *too clean* and that certain levels of microbial exposure are fundamental to human health [60,61]. This theory, known as the hygiene hypothesis, currently represents one of the most popular explanations for the dramatic increases in childhood illnesses and resistance to antibiotics [57,62]. In medical and scientific discourse, there is a growing interest in the microbial theory of health, which promotes *targeted* hygiene to manage microbial diversity, balancing exposure to naturally diverse ecological systems in place of sterilization measures. Instead of focusing on indiscriminate sterilization procedures, this approach emphasizes the use of quantitative tools to understand the complex ecological relationships that link hygienic behavior to the spread of disease [63]. Childhood exposure to dogs, for example, is associated with a decreased risk of allergies and asthma [64]. This probiotic, 21st-century approach to health is hoped to build upon germ theory and provoke a new era in which pathogen reduction is done without indiscriminately eliminating naturally occurring and potentially beneficial microorganisms from the human and environmental microbiomes [62].

Evidence suggests that we must (again) critically question the beneficial relationships between the built environment and human health. By recalibrating public understanding of hygiene and what defines a healthy and equitable space, it is possible to rethink the paradigm of how we design and occupy buildings. By factoring in the healthful effects of fresh air, architectural space, and the indoor microbiome, this book provides strategies to promote human health in buildings, blurring the edges between inside and out. Many of the modern habits associated with the built environment and mechanical conditioning technologies stem from psychological associations that are culturally constructed. Though the health benefits of access to fresh air are easily quantified, designing for humans is inherently complex. The next chapter will begin to address the socio-cultural factors pushing humans so deeply indoors.

Notes

i On January 6th, 2023, the EPA announced that it would revise the annual $PM_{2.5}$ standard from its current level of 12.0 µg/m [3] to within the range of 9–10 µg/m [39]. However, this standard is only for ambient conditions and still does not regulate indoor air.

ii This proportion varies by state: Illinois 21.1%, California 20.1%, New York 18.8%, compared against Ohio 9.5% and Florida 3% [18].

References

[1] Climate & Clean Air Coalition, dir. 2016. "Dr Margaret Chan, WHO Director General, Message to the UN Environment Assembly." www.youtube.com/watch?v=tQwXjvQsQ4I.

[2] Holgate, Stephen T. 2017. " 'Every Breath We Take: The Lifelong Impact of Air Pollution' – a Call for Action." *Clinical Medicine* 17 (1): 8–12. https://doi.org/10.7861/clinmedicine.17-1-8.

[3] Bruce, N., R. Perez-Padilla, and R. Albalak. 2000. "Indoor Air Pollution in Developing Countries: A Major Environmental and Public Health Challenge." *Bulletin of the World Health Organization* 78 (9): 1078–92.

[4] US Environmental Protection Agency. 1987. *The Total Exposure Assessment Methodology (TEAM) Study*. Washington, DC: Office of Research and Development.

[5] González-Martín, Javier, Norbertus Johannes Richardus Kraakman, Cristina Pérez, Raquel Lebrero, and Raúl Muñoz. 2021. "A State–of–the-Art Review on Indoor Air Pollution and Strategies for Indoor Air Pollution Control." *Chemosphere* 262 (January): 128376. https://doi.org/10.1016/j.chemosphere.2020.128376.

[6] Jones, A. P. 1999. "Indoor Air Quality and Health." *Atmospheric Environment* 33 (28): 4535–64. https://doi.org/10.1016/S1352-2310(99)00272-1.

[7] Sundell, J. 2004. "On the History of Indoor Air Quality and Health." *Indoor Air* 14 (s7): 51–8. https://doi.org/10.1111/j.1600-0668.2004.00273.x.

[8] "Dementia Risk May Rise as Air Quality Worsens." April 5, 2023. www.medpagetoday.com/neurology/dementia/103884.

[9] US Environmental Protection Agency. 2023. "EPA Proposes to Strengthen Air Quality Standards to Protect the Public from Harmful Effects of Soot." January 6. www.epa.gov/newsreleases/epa-proposes-strengthen-air-quality-standards-protect-public-harmful-effects-soot.

[10] Kraus, Michal. 2016. "Airtightness as a Key Factor of Sick Building Syndrome (SBS)." In *International Multidisciplinary Scientific GeoConference: SGEM*, vol. 2, 439–45. Sofia, Bulgaria: Surveying Geology & Mining Ecology Management (SGEM). www.proquest.com/docview/2014512056/abstract/66CDBCEAD4714D68PQ/1.

[11] Joshi, Sumedha M. 2008. "The Sick Building Syndrome." *Indian Journal of Occupational and Environmental Medicine* 12 (2): 61–4. https://doi.org/10.4103/0019-5278.43262.

[12] Rios, José Luiz de Magalhães, José Laerte Boechat, Adriana Gioda, Celeste Yara dos Santos, Francisco Radler de Aquino Neto, and José Roberto Lapa e Silva. 2009. "Symptoms Prevalence among Office Workers of a Sealed versus a Non-Sealed Building: Associations to Indoor Air Quality." *Environment International* 35 (8): 1136–41. https://doi.org/10.1016/j.envint.2009.07.005.

[13] US Environmental Protection Agency. 1989. *Report to Congress on Indoor Air Quality, Volume II: Assessment and Control of Indoor Air Pollution.* Technical Report EPA/400/1-89/001C. Washington, DC: USEPA.

[14] Awada, Mohamad, Burcin Becerik-Gerber, Simi Hoque, Zheng O'Neill, Giulia Pedrielli, Jin Wen, and Teresa Wu. 2021. "Ten Questions Concerning Occupant Health in Buildings during Normal Operations and Extreme Events Including the COVID-19 Pandemic." *Building and Environment* 188 (January): 107480. https://doi.org/10.1016/j.buildenv.2020.107480.

[15] Fisk, William J., and Arthur H. Rosenfeld. 1997. "Estimates of Improved Productivity and Health from Better Indoor Environments." *Indoor Air* 7 (3): 158–72. https://doi.org/10.1111/j.1600-0668.1997.t01-1-00002.x.

[16] Shrestha, Prateek M., Jamie L. Humphrey, Elizabeth J. Carlton, John L. Adgate, Kelsey E. Barton, Elisabeth D. Root, and Shelly L. Miller. 2019. "Impact of Outdoor Air Pollution on Indoor Air Quality in Low-Income Homes during Wildfire Seasons." *International Journal of Environmental Research and Public Health* 16 (19): 3535. https://doi.org/10.3390/ijerph16193535.

[17] Lin, Weiwei, Bert Brunekreef, and Ulrike Gehring. 2013. "Meta-Analysis of the Effects of Indoor Nitrogen Dioxide and Gas Cooking on Asthma and Wheeze in Children." *International Journal of Epidemiology* 42 (6): 1724–37. https://doi.org/10.1093/ije/dyt150.

[18] Gruenwald, Talor, Brady A. Seals, Luke D. Knibbs, and H. Dean Hosgood. 2023. "Population Attributable Fraction of Gas Stoves and Childhood Asthma in the United States." *International Journal of Environmental Research and Public Health* 20 (1): 75. https://doi.org/10.3390/ijerph20010075.

[19] American Society of Heating and Ventilating Engineers. 1895. *Transactions of the American Society of Heating and Ventilating Engineers.* The Society.

[20] Yaglou, C. P., E. C. Riley, and D. I. Coggins. 1936. *Ventilation Requirements*, vol. 42, 1031. Boston, MA: ASHRAE. www.aivc.org/resource/ventilation-requirements-0.

[21] Persily, Andrew. 2015. "Challenges in Developing Ventilation and Indoor Air Quality Standards: The Story of ASHRAE Standard 62." *Building and Environment, Fifty Year Anniversary for Building and Environment* 91 (September): 61–9. https://doi.org/10.1016/j.buildenv.2015.02.026.

[22] American Society of Heating, Refrigerating and Air-Conditioning Engineers Inc Standards Committee January 28 approved by the ASHRAE Standard. 1973. *Standards for Natural and Mechanical Ventilation*, 62–73. New York: ASHRAE.

[23] Ackermann, Marsha. 2002. *Cool Comfort: America's Romance with Air-Conditioning.* Washington, DC: Smithsonian Institution.

[24] Allen, Joseph G., Piers MacNaughton, Usha Satish, Suresh Santanam, Jose Vallarino, and John D. Spengler. 2016. "Associations of Cognitive Function Scores with Carbon Dioxide, Ventilation, and Volatile Organic Compound Exposures in Office Workers: A Controlled Exposure Study of Green and Conventional Office Environments." *Environmental Health Perspectives* 124 (6): 805–12. https://doi.org/10.1289/ehp.1510037.

[25] MacNaughton, Piers, James Pegues, Usha Satish, Suresh Santanam, John Spengler, and Joseph Allen. 2015. "Economic, Environmental and Health Implications of Enhanced Ventilation in Office Buildings." *International Journal of Environmental Research and Public Health* 12 (11): 14709–22. https://doi.org/10.3390/ijerph121114709.

[26] Fadaei, Abdolmajid. 2021. "Ventilation Systems and COVID-19 Spread: Evidence from a Systematic Review Study." *European Journal of Sustainable Development Research* 5 (2): em0157. https://doi.org/10.21601/ejosdr/10845.

[27] Guo, Mingyue, Peng Xu, Tong Xiao, Ruikai He, Mingkun Dai, and Shelly L. Miller. 2021. "Review and Comparison of HVAC Operation Guidelines in Different Countries During the COVID-19 Pandemic." *Building and Environment* 187 (January): 107368. https://doi.org/10.1016/j.buildenv.2020.107368.

[28] Janssen, John E. 1999. "The History of Ventilation and Temperature Control: The First Century of Air Conditioning." *ASHRAE Journal* 41 (10): 48.

[29] Klauss, A. K., R. H. Tull, L. M. Roots, and J. R. Pfafflin. 2011. "History of the Changing Concepts in Ventilation Requirements." *ASHRAE Journal* 53 (2): 34–6, 38, 40, E42–3.

[30] Hippocrates. 1886. *The Genuine Works of Hippocrates*. Wood.

[31] Collins, Julie. 2012. "Life in the Open Air: Place as a Therapeutic and Preventative Instrument in Australia's Early Open-Air Tuberculosis Sanatoria." *Fabrications* 22 (2): 208–31. https://doi.org/10.1080/10331867.2012.733161.

[32] Tountas, Yannis. 2009. "The Historical Origins of the Basic Concepts of Health Promotion and Education: The Role of Ancient Greek Philosophy and Medicine." *Health Promotion International* 24 (2): 185–92. https://doi.org/10.1093/heapro/dap006.

[33] Fischer, Louis. 1925. *The Health-Care of the Baby: A Handbook for Mothers and Nurses*. New York: Funk & Wagnalls.

[34] Atkinson, V. Sue. 2022. "Every Picture Tells a Story: Parenting Advice Books Provide a Window on the Past." *Social Sciences* 11 (1): 11. https://doi.org/10.3390/socsci11010011.

[35] Holt, Luther Emmett. 1909. *The Care and Feeding of Children*. New York: D. Appleton.

[36] U.S. Department of Labor, Children's Bureau. 1935. *Infant Care*. Bureau Publication No. 8. Washington, DC: Government Printing Office. www.mchlibrary.org/history/chbu/3121-1935.PDF.

[37] Read, Emma. 1923. "Portable Baby Cage, Issued 1923." https://worldwide.espace-net.com/publicationDetails/biblio?FT=D&date=19230313&DB=EPODOC&CC=US&NR=1448235A.

[38] Kushnick, Geoff. 2019. "The Cradle of Humankind: Evolutionary Approaches to Technology and Parenting." *SocArXiv* 115. https://doi.org/10.31235/osf.io/k23md.

[39] Keyser, Hannah. 2015. "A Brief and Bizarre History of the Baby Cage." June 24. www.mentalfloss.com/article/65496/brief-and-bizarre-history-baby-cage.

[40] Gilbert, Heather A. 2020. "Florence Nightingale's Environmental Theory and Its Influence on Contemporary Infection Control." *Collegian, Special Issue on Nursing and Health Care History* 27 (6): 626–33. https://doi.org/10.1016/j.colegn.2020.09.006.

[41] Medeiros, Ana Beatriz de Almeida, Bertha Cruz Enders, and Ana Luisa Brandão De Carvalho Lira. 2015. "The Florence Nightingale's Environmental Theory: A Critical Analysis." *Escola Anna Nery* 19 (September): 518–24. https://doi.org/10.5935/1414-8145.20150069.

[42] Damsky, Ellen. 2003. "A Way of Life: Saranac Lake and the 'Fresh Air' Cure for Tuberculosis." Ph.D., State University of New York, Binghamton. www.proquest.com/docview/305238703/abstract/7E0A9CDECAB74D42PQ/1.

[43] CDC. 2021. "Tuberculosis (TB) – World TB Day – History." Centers for Disease Control and Prevention. January 28, 2021. www.cdc.gov/tb/worldtbday/history.htm.

[44] Murray, John F., Dean E. Schraufnagel, and Philip C. Hopewell. 2015. "Treatment of Tuberculosis. A Historical Perspective." *Annals of the American Thoracic Society* 12 (12): 1749–59. https://doi.org/10.1513/AnnalsATS.201509-632PS.

[45] Campbell, Margaret. 2005. "What Tuberculosis Did for Modernism: The Influence of a Curative Environment on Modernist Design and Architecture." *Medical History* 49 (4): 463–88.

[46] Hobday, Richard A., and John W. Cason. 2009. "The Open-Air Treatment of Pandemic Influenza." *American Journal of Public Health* 99 (S2): S236–42. https://doi.org/10.2105/AJPH.2008.134627.

[47] Greenhalgh, I., and A. R. Butler. 2017. "Sanatoria Revisited: Sunlight and Health." *The Journal of the Royal College of Physicians of Edinburgh* 47 (3): 276–80. https://doi.org/10.4997/jrcpe.2017.314.

[48] Greene, Gina. 2011. "Nature, Architecture, National Regeneration: The Airing Out of French Youth in Open-Air Schools 1918–1939." Working Paper 1362. Princeton University, School of Public and International Affairs, Center for Arts and Cultural Policy Studies. https://econpapers.repec.org/paper/pricpanda/45.htm.

[49] "International Congress on Hygiene and Demography: Fifteenth Meeting, Held at Washington, DC, September 23–28, 1912." 1912. *Journal of the American Medical Association* LIX (16): 1480–84. https://doi.org/10.1001/jama.1912.04270100248031.

[50] Greene, Gina Marie. 2012. "Children in Glass Houses: Toward a Hygienic, Eugenic Architecture for Children during the Third Republic in France (1870–1940)." Ph.D., Princeton University, Princeton. Accessed April 8, 2023. www.proquest.com/docview/1240667687/abstract/4636B9DE1C3D4C18PQ/1.

[51] Korr, Mary. 2016. "Fighting TB with Fresh-Air Schools." *Rhode Island Medical Journal* 99 (9): 75–6.

[52] Nierenberg, Amelia. 2020. "Classrooms Without Walls, and Hopefully Covid." *The New York Times*, October 27, sec. U.S. www.nytimes.com/2020/10/27/us/outdoor-classroom-design.html.

[53] Bauld, Andrew. 2021. "Make Outdoor Learning Your Plan A. Usable Knowledge." *Harvard Graduate School of Education* (blog), August 18. www.gse.harvard.edu/news/uk/21/08/make-outdoor-learning-your-plan.

[54] Will, Madeline. 2021. "If Outdoor Learning Is Safer During COVID, Why Aren't More Schools Doing It?" *Education Week*, September 14, sec. Teaching & Learning, Teaching. www.edweek.org/teaching-learning/if-outdoor-learning-is-safer-during-covid-why-arent-more-schools-doing-it/2021/09.

[55] Ramirez-Figueroa, Carolina, and Richard Beckett. 2020. "Living with Buildings, Living with Microbes: Probiosis and Architecture." *ARQ: Architectural Research Quarterly* 24 (2): 155–68. https://doi.org/10.1017/S1359135520000202.

[56] Hobday, Richard A. 2019. "The Open-Air Factor and Infection Control." *Journal of Hospital Infection* 103 (1): e23–24. https://doi.org/10.1016/j.jhin.2019.04.003.

[57] Blaser, Martin J. 2015. *Missing Microbes: How the Overuse of Antibiotics Is Fueling Our Modern Plagues*. First Picador edition. New York: Picador. https://catalog.lib.ncsu.edu/catalog/NCSU3511142.

[58] Dancer, Stephanie J. 2013. "Infection Control in the Post-Antibiotic Era." *Healthcare Infection* 18 (2): 51–60. https://doi.org/10.1071/HI12042.

[59] Beckett, Richard. 2021. "Probiotic Design." *The Journal of Architecture* 26 (1): 6–31. https://doi.org/10.1080/13602365.2021.1880822.

[60] Greenhough, Beth, Andrew Dwyer, Richard Grenyer, Timothy Hodgetts, Carmen McLeod, and Jamie Lorimer. 2018. "Unsettling Antibiosis: How Might Interdisciplinary Researchers Generate a Feeling for the Microbiome and to What Effect?" *Palgrave Communications* 4 (1): 1–12. https://doi.org/10.1057/s41599-018-0196-3.

[61] Cirstea, Mihai, Nina Radisavljevic, and B. Brett Finlay. 2018. "Good Bug, Bad Bug: Breaking Through Microbial Stereotypes." *Cell Host & Microbe* 23 (1): 10–13. https://doi.org/10.1016/j.chom.2017.12.008.

[62] Scott, Elizabeth A., Elizabeth Bruning, Raymond W. Nims, Joseph R. Rubino, and Mohammad Khalid Ijaz. 2020. "A 21st Century View of Infection Control in Everyday Settings: Moving from the Germ Theory of Disease to the Microbial Theory of Health." *American Journal of Infection Control* 48 (11): 1387–92. https://doi.org/10.1016/j.ajic.2020.05.012.

[63] Vandegrift, Roo, Ashley C. Bateman, Kyla N. Siemens, May Nguyen, Hannah E. Wilson, Jessica L. Green, Kevin G. Van Den Wymelenberg, and Roxana J. Hickey. 2017. "Cleanliness in Context: Reconciling Hygiene with a Modern Microbial Perspective." *Microbiome* 5 (1): 76. https://doi.org/10.1186/s40168-017-0294-2.

[64] Horve, Patrick F., Savanna Lloyd, Gwynne A. Mhuireach, Leslie Dietz, Mark Fretz, Georgia MacCrone, Kevin Van Den Wymelenberg, and Suzanne L. Ishaq. 2020. "Building upon Current Knowledge and Techniques of Indoor Microbiology to Construct the Next Era of Theory into Microorganisms, Health, and the Built Environment." *Journal of Exposure Science & Environmental Epidemiology* 30 (2): 219–35. https://doi.org/10.1038/s41370-019-0157-y.

2 The Socially Constructed Microbiome

Elizabeth L. McCormick

In 1972, a musician from the San Francisco Symphony Orchestra recorded a violin solo in the back of a '72 Ford LTD to demonstrate the pristine silence of the interior cabin. "Take a quiet break in the Ford LTD", a man's voice says in the ad, "where quiet plus a luxury ride means a better car, a better value for you" [1]. Nearly 40 years later, a man is locked inside a Toyota Corolla in a 2008 Superbowl commercial, with a sleeping mother badger nursing her pups in the passenger seat beside him. "If awakened, they'll gnaw his face off", a narrator warns as the characters outside of the vehicle proceed to launch cannons. With the sound-isolating capabilities of the Corolla on full display, the badger family continues to sleep peacefully . . . until the man's cell phone rings inside the car, awakening the mother badger. The commercial cuts to black and over the sounds of what we can only assume to be a vicious badger attack, we hear, "The more refined Corolla. Live the dream for less coin" [2]. Throughout recent history, the automobile, a triumph of human engineering, has been marketed as a retreat from the noise and chaos of the outside world – the more isolated, the better. Though advertising material is meant to be quippy and memorable, this is a common sentiment in modern culture, where the outdoors is portrayed as turbulent, noisy, and unclean, while the engineered interior is calm and collected, a testament to human control and mastery over nature.

> The opposites of wildness and animality were civilization and humanity . . . Outside the boundary was disorderly wilderness, inside ordered civilization . . . To be civilized was to impose order on personal life; civilization represented the imposition of order on the land.
> —Carolyn Merchant, *Reinventing Eden*, 2003. [3]

In *The Experience of Landscape* (1975), Jay Appleton describes an innate human need, that of *refuge* and *prospect*. According to his theory, this psychological condition stems from a primitive need to view one's prey without being spotted themselves. Though for the most part we are no longer hunters and gatherers, Appleton believes that this explains why the spaces where we feel most

DOI: 10.1201/9781003398714-3

comfortable require sensitivity to both opportunity and security [4], a calming mix of opposing conditions. In fact, evidence shows that the human body is most satisfied in mixed, heterogeneous environments – both thermally and spatially [5–9]. Modern mechanical engineering, however, is built on the premise that environmental stimuli should be nullified to produce reliable and consistent levels of human comfort, independent from the caprice of weather and climate. Despite the thermal conditions provided indoors, there are a number of factors influencing people's perceptions of indoor space that are not technological or quantifiable in nature. In fact, many of the drivers pushing people indoors are socially and culturally constructed and not based on scientific reasoning or biological intuitions.

> Tools always presuppose a machine, and the machine is always social before it is technical. There is always a social machine which selects or assigns the technical elements used.
>
> —Kiel Moe, *Compelling Yet Unreliable Theories of Sustainability*, 2007 [10]

The human relationship with dirt, germs, and cleanliness is actually a relatively new construct that stems from social theories of health and hygiene. Because of the close associations between sociocultural constructs and technological development, truly healthy buildings must embrace both technology and human behavior concurrently. In support of this effort, this chapter will address the social constructs that have influenced our understanding (and use of) of indoor space through three primary human behaviors: control, fear, and technocratic utopianism.

Climate Control

Outdoor air has a tainted history through the industrialized West, driven by sociocultural reaction to the blights of urban phenomena. Distaste for outdoor air grew from fears of health effects of respiratory urban diseases such as tuberculosis and pneumonia, as well as coal-driven pollution. However, the modernization of the built environment through mechanical conditioning technologies catalyzed a profound transformation of air "from menace to modernizing agent" [11]. *The Aerologist*, the first trade journal on air conditioning, predicted that air conditioning would become indispensable to American life – more so than the radio or automobile. On the cover of the June 1929 issue, shown in Figure 2.1, a woman and her child are protected within a windowless, artificially lit domestic environment while a wall separates her from the toxic outdoors where a man in business attire dons a gas mask. Increasingly, mechanically conditioned indoor environments were flaunted as safe and healthy while the outdoors were perceived as

Figure 2.1 "A Fresh Air Fan of 1935". The Aerologist, Magazine Cover, June 1929.

increasingly toxic and unpredictable. Psychologically, post-war Americans defined interior space through stark isolation, controlled and predictable. Though pre-air-conditioned buildings were intimately connected to a region's local climate, the role of the built environment was increasingly to protect its occupants from nature.

The view that manmade technology promised a more prosperous future permeated American culture in the early 1900s, and the ability to control a building's interior environment became a symbol of independence from the forces of nature, creating a sense of human autonomy in an increasingly techno-centric population. The commonly cited quip that "everyone talks about the weather, but no one does anything about it", from American writer Charles Dudley Warner (though it is often attributed to Mark Twain), was used to express the futility of trying to modify things that are impossible to change. Unlike nature, machines could provide manufactured environments that were controlled and routine. Gail Cooper, technological historian and author of *Air-Conditioning America*, details the history of conditioned air, which the Carrier Corporation described as 'Manmade weather' until the term 'air-conditioning' was introduced in the 1930s. In its earliest stages, this technology was utilized mostly for industrial purposes, such as in factories to ensure smooth operation of the equipment, which was also becoming increasingly mechanized. *Weatherlessness* and *manufactured weather* were promoted as tenets of modernism and engineers flaunted the idea, among other technical marvels of the modern era, of mastery over nature [12]. This *modern air* required custom-made environments that necessitated complete enclosure and separation from the outdoors. The idea of artificial environments and the "technocratic vision of perfect man-made weather" was particularly alluring to an increasing population of consumers who welcomed the escape from climate that air conditioning provided [12].

Like the car commercials mentioned at the beginning of this chapter, this notion of demonizing the outdoors as unsafe and anti-modern was reflected in popular culture as well. Edward Bellamy's bestselling 1887 novel, *Looking Backward*, outlined a utopian and socialist vision for Boston in the year 2000, where everyday tasks were conducted by machines, in stark contrast to the technological climate in which it was written. Bellamy describes the "extraordinary imbecility to permit the weather to have any effect on the social movements of people" [13,14]. Though his high-tech vision of the future was written as fiction, it represented emerging social beliefs impacting modern perceptions of conditioned indoor space. Nearly 50 years later, H.G. Wells' 1933 science fiction novel, *The Shape of Things to Come*, tells the story of a Europe devastated by plague that rebuilds as a technocratic civilization, solving all of humanity's problems with machines. The 1936 movie rendition, *Things to Come*, was saturated with visuals of what was hoped for future civilization. In one scene, shown in Figure 2.2, a man is showing 'history pictures' to his young great-granddaughter. They pause on a photograph of New York . . .

Figure 2.2 Still from H.G. Wells' "Things to Come" [1:14:36], 1936.

Girl: What a funny place New York was, all sticking up and full of windows.
Man: They built houses like that in the old days
        Why?
        They'd no light inside their cities as we have, so they'd have to stick
        them up into the daylight . . . what there was of it. They had no properly
        mixed and conditioned air. Everybody lived half out of doors. They had
        windows of brittle glass. The Age of Windows lasted four centuries – they
        never seemed to realize that we could light the interiors of our houses with
        sunshine of our own, so there was no need to stick them up ever so high
        in the air.
        Weren't people tired of going up and down those stairs?
        They were all tired, and they had a disease called colds. Everybody had
        colds. And they coughed and sneezed and ran at the eyes.
        Sneezed? What's 'sneezed'?
        Oh, you know, 'Atishoo!' (the man makes a sneezing gesture)
        Atishoo? Everyone said "atishoo"? That must have been funny
        Not as funny as you think
        And you remember all of that, great-grandfather?
        Well, I remember some of it. Colds we had, and indigestion too from
        the queer bad foods we ate. Oh, it was a poor life. Never really well.

Did people laugh at it?
They had a way of grinning at it. They used to call it "humor". We had to have a lot of humor. I've lived through some horrid times, my dear. Oh, horrid.
Horrid. I don't want to hear about that (girl leaves his side and walks away) [15]

Wells was not alone in his vision of windowless, artificial environments. Early air conditioning contracts in industrial settings actually required that windows remain closed. The *Southern Textile Bulletin* (Clark Publishing), which began publication in 1911, often contained anti-window propaganda, as seen in these Parks-Cramer Co. advertisements (first referenced in Cooper 2002) from the 1936 *Textile Bulletin*, which was one of the most widely distributed industrial journals to owners and managers of mills in the South. In Figure 2.3, a bandit climbs in through an open window next to a caption that reads, "It's an ill wind that blows through open windows", inflating the health fears associated with outdoor air. The advertisement in Figure 2.4 responds to the idea of lost profit from environmental instability, stating "DOWN with windows UP with Profits". Mechanized climate control (and closed windows) meant that businesses could operate efficiently at all times of the year, regardless of the weather. At one point, a region's climate influenced the local way of life. In the hot and humid South, for example, people would adopt a slower pace during the summer. Instead, air conditioning (and closed windows) meant that climate would no longer dictate a region's economic (or social) capacities throughout the year.

Large windowless department stores emerged in cities across the United States. Foley's department store opened in downtown Houston, Texas, in 1947 and was promoted as "the cleanest in the world" [16] and "another triumph of air conditioning" [17]. Even in residential buildings, mechanical ventilation took over the role of open windows in the 1950s and 60s. By extension, modern conditioning brought a shift from human-centered design to spaces that responded first to technical limitations [12], using technology to "liberate humanity from all the ills of the human condition" [18]. The adoption of mechanical conditioning into the American home in the 1960s fermented the link between climate control and social status in the United States [19]. It soon became unacceptable for people to sweat or smell of body odor, for example, creating a demand for air conditioning in all interior environments, regardless of climate. Consequently, people developed specific interior comfort expectations and their original perceptions of thermal satisfaction were forever changed [20].

In a true demonstration of the triumph of manufactured weather, Houston was dubbed "the most air-conditioned city in the world" in 1955 [21]. Despite Houston's extreme heat and humidity, athletes and spectators alike enjoyed a consistent 74 degrees and 50% humidity in the Houston Astrodome, the first air-conditioned athletic stadium. Known as the Eighth Wonder of the World, the domed structure represented "the city's absolute mastery of the region's environment", and

Figure 2.3 Parks-Cramer Co. Advertisement. Southern Textile Bulletin, 1936.

"spectators regularly found anything from a hot dog and a cold beer to lobster and hot toddies as they relaxed alongside the dome's $4.5 million cooling system" [16,21]. In the late 1960s, the Houston Galleria was constructed as an archetype for future malls across America. Not only did this commercial megaplex

Figure 2.4 Parks-Cramer Co. Advertisement. Southern Textile Bulletin, 1936.

contain endless halls of conditioned air despite summer conditions often in excess of 95 degrees; it also has a year-round ice-skating rink. And with miles of underground tunnels connecting downtown office buildings, it could also be argued that Houston is the city that greatest represents the disparate conditions between inside and out, both culturally and climatically.

Viewed through a deterministic lens, the development of machines forever changed the human social fabric and our relationship with nature. Eminent architectural critic Reyner Banham published a comprehensive history in *The Architecture of the Well-Tempered Environment*, where he documented the influence of "technological art" in the built environment. Banham believed that 1882, the year that electric power was domesticated, set the foundation for more sophisticated environmental advances, such as the control of humidity (the cornerstone of modern air-conditioning systems). It was at this point that the responsibility of environmental control was given to *systems*, not *structure*, and "the possibility of absolute variety and infinite choice of building form is now with us" [22]. Post-war modernism was largely influenced by a *machine-made* aesthetic [23] and driven by a utopian vision of the transformative possibilities of technology in a new scientific view – using nature's principles to achieve what evolution could not [24]. The International Style, for example, was an architectural response to industry, technology, mobility, and sociopolitical orders with an aesthetic that was austere and abstract, founded on the ideas of universalism and mass production [25,26]. Concern for passive design strategies like orientation, shading, site planning, and ventilation diminished rapidly as air conditioning was widely adopted across the United States, and the futuristic notion of buildings that were entirely controlled by mechanically conditioned air found fertile ground amid a growing technological enthusiasm in the United States [11].

Dirt, Germs, and Hygiene

During the Renaissance (14–17th centuries), etiquette – not sanitation – was the primary driver of hygiene practices. Spitting, coughing, and sneezing, for example, were considered beastlike behaviors that had to be concealed among those seeking social distinction and civility. This idea, not the fear of disease, drove a code of manners rooted in gentility and politeness that began within the social elite and diffused through the urban middle class [27]. Though it may seem obvious now that disease is spread through direct contact, humans were mainly living in rural settings with generous access to fresh air until the Industrial Revolution. It wasn't until urban populations densified that the transmission of disease between humans became more apparent [28]. Until the development of microbiology in the mid-19th century, people developed a wide range of theories to explain the transmission of infectious diseases. The miasma theory, for example, was the dominant view among scientists and doctors, which proposed that what we now identify as communicable diseases were caused by *miasma*, or poisonous vapors found in 'bad air' [29,30]. Other diseases were often thought to come from poor dispositions, from unhealthy living habits, or perceived inferior genetics [27]. Tuberculosis was once thought by scientists to be a social disease rooted in the 'evils' of urban inhabitants, particularly immigrants [31].

According to social anthropologist Mary Douglas, nature has been portrayed throughout history as dirty, symbolic of disorder, a lack of sophistication, immoral ways of living, and even danger [32,33]. Though the disgust of dirt has biological roots [34], the association between dirt and disease wasn't formed until the cholera epidemic of 1832. In fact, historians and anthropologists believe that concern for dirt and filth is a relatively new and fundamentally social construct, with different perceptions of what counts as 'dirty' based on social norms, gender, class, and numerous other factors [27,35]. As late as the 19th century, dirt was actually considered a virtue by rural Americans, a symbol of austerity, hard work, and physical strength [36]. Baths were irregular, taken once a week before church, and rarely included soap, which dried the skin. Suellen Hoy's 1996 book, *Chasing Dirt: The American Pursuit of Cleanliness* was the first general history of cleanliness in the United States and is an excellent source of information about the cultural underpinnings of hygienic practices in the United States. Similarly, Mary Douglas' 1966 book, *Purity and Danger*, is viewed as a key text of social anthropology and is a pivotal source for this narrative. In pursuing a clean, dirt-free environment, Douglas believes that humans are responding to the disorder and unpredictability of nature in their quest for a world that is more structured around them. Dirt was a fundamental disturbance that had to be purged to maintain the integrity of the system.

> Dirt offends against order. Eliminating it is not a negative movement, but a positive effort to organize the environment . . . In chasing dirt, in papering, decorating, tidying, we are not governed by anxiety to escape disease, but are positively re-ordering our environment, making it conform to an idea. There is nothing fearful or unreasoning in our dirt-avoidance: it is a creative movement, an attempt to relate form to function, to make unity of experience.
> —Mary Douglas, *Purity and Danger* [32]

The notion that dirt was associated with filth and the evils of the natural world was evident in religious texts as well. The Bible states, "Wash me clean of my guilt, purify me from my sin" (Psalms 51:2), and the Koran says, "God loves those that turn to him in repentance and strive to keep themselves clean" (2:223) [34]. These ideas permeated domestic life, particularly for women, who were the primary keepers of the home. Ellen Richards, a pioneer of the science of home economics in the late 19th century, declared that "dirt is a sin" and that women's important work within the home supported both cleanliness *and* morality [27]. The eminent link between disease and domestic hygiene ultimately expanded women's obligations within the home, making housekeeping both a physical and psychological chore. Figure 2.5 shows a woman sweeping away dust to combat the threats of typhoid fever, consumption, influenza, germs, and microbes, each with their own corresponding cloud. A shrouded figure representing Death observes from the background, rubbing his hands with approval. In addition to caring for the home, women sought to provide safe, hygienic lifestyles for families.

Ultimately, it wasn't until the mid-1800s that scientists learned that diseases were caused by microorganisms, particularly bacteria, forming the underpinnings of germ theory. This was a practical concept that was easily understood and

Figure 2.5 Illustration showing communicable diseases spread by household and street dust, 1920. Image provided courtesy of the National Library of Medicine.

quickly embraced by the general public. With the use of the microscope, germs were visible to the human eye. Because of that, they were easy to blame for so much damage, a critic noted in *Popular Science Monthly* in 1885 [27]. Visuals of the 'evil' germ began to permeate popular culture. In *Giro the Germ* (1927), a short cartoon produced by the Health and Cleanliness Council, germs were portrayed as tiny black monsters that travel along the feet of the flies attracted to unkempt homes, shown in Figure 2.6. A voice reads . . .

> See Where Giro's Going, Flying Through the Air? Where There's Dirt, There's Giro, and Danger Everywhere! [37]

Giro was part of a larger wartime health campaign in Europe that promoted cleanliness in the purported service of national security. In *See How They Won* (1935), a cartoon commissioned by the British drug store Boots Chemist, an unassuming gentleman named *John Careless* is the target of a military-like squad

Figure 2.6 Still from *Giro the Germ* [1:45], 1927.

of germs. As poor John leaves his house to go to work, a catchy but ominous jingle plays . . .

> *John Careless leaves his home each morn,*
> *the rules of healthy treats bescorn,*
> *never been ill since he was born.*
>
> *His home's quite unprotected,*
> *But there are germs so gray and grim*
> *who lie in wait for folks like him,*
> *and hope someday to do them in when they least expect it.* [38]

The army of cartoon characters then pours a jar of 'germs' into his office fan to infect him, shown in Figure 2.7. Though these are just two examples of the representation of disease-causing microorganisms, numerous strategies were found to reimagine germs to reflect societal perceptions of disgust, worry, and concern for their homes and workplaces [39]. The figures were given relatable human qualities, which were soon replaced with increasing degrees of scientific integrity after the Second World War, as shown in Figure 2.8.

By the early 1900s, America was increasingly becoming a consumer-driven society due to industrial progress and modernization. With the influx of televisions into American life in the 1940s and 50s, visual representations of dirt, germs, and hygiene practices were becoming more visual, often enlisting familiar

Figure 2.7 Still from *See How They Won* [1:50], 1935.

prejudices and anxieties related to gender, class, race, and ethnicity. This emerging visual culture and popularization of germ aversion presented opportunities for companies to promote sales through advertising in newspapers and magazines. Many different manufacturers quickly capitalized on public health concerns, sensationalizing risk in order to attract attention and promote sales [40]. As a result, advertising played a vital role in domesticating and normalizing technologies that seemed frightening at first [14]. Advertisements for hand soaps and detergents, for example, underscored the fear of germs, while advertisements for prescription drugs also emphasized visual representations of germs [39].

Ultimately, the realization that diseases could be spread through human transmission, not environmental conditions, induced a culture of excessive cleanliness and urban germaphobia. Most notably, it triggered what American historian Nancy Tomes describes as a 'gospel of germs' – the belief that disease could be avoided by certain protective behaviors [27]. Though it did take some time for middle-class Americans to comprehend that they could take practical steps to prevent the spread of disease, the powerful association between dirt, filth, and disease permeated every aspect of American life [36]. Capitalizing on this growing fear, American soap manufacturers launched the Cleanliness Institute in 1927 to educate the public on the virtues of cleanliness. This group focused primarily on school-age children, developing a 'cleanliness curriculum' to enforce hygienic behaviors which, of course, involved the generous use of soap products [41]. In Britain, the Health and Cleanliness Council (HCC) promoted similar hygiene-focused educational materials, including Giro the Germ, mentioned earlier. Funded by soap and electrical industries, HCC promoted a number

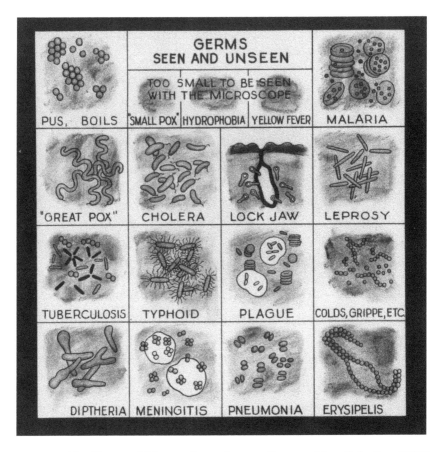

Figure 2.8 The Germs of Disease, date unknown. Keystone-Mast Collection, UCR/ California Museum of Photography, University of California at Riverside.

of domestic electrical technologies, such as artificial lighting [42]. In a 1934 address, the first president and chairman of HCC advised that houses should be constructed so that they were easily kept clean, simply maintained, and with provisions for the use of electricity to both save time and improve cleanliness [39]. Both organizations, the Cleanliness Institute and Health and the Health and Cleanliness Council, exploited (and even helped to promote) the fear of dirt and filth that was rapidly spreading through Western culture. In fact, the HCC's motto was, "Where there's dirt there's danger", as shown in Figure 2.9.

In the 1920s and 30s, self-service grocery stores emerged rapidly in the United States, made possible by emerging refrigeration technologies and disposable, single-use packaging products, for which consumers grew accustomed to paying for in addition to the goods. Many other disposable products emerged during the same period, claiming to support a more sanitary, germ-free lifestyle. Kleenex, for example, was introduced in 1924 as a cold cream remover. However, when

Figure 2.9 Logo of the Health and Cleanliness Council, 1937. Archive of the National Association of Teachers of Home Economics and Technology, earlier the Association of Teachers of Domestic Science, Modern Records Centre, University of Warwick. Originally cited in Greenway, 2019.

it was found to be more commonly used for nose-blowing, the company quickly adjusted the marketing material toward cleanliness, portraying cloth handkerchiefs as both unsanitary and uncivilized, linking hygiene practices with perceptions of class and sophistication. Protective behaviors became increasingly evident in a consumer culture when marketers targeted women, the demographic most concerned with the implications of domestic cleanliness.

DuPont, the maker of Cellophane, launched an aggressive and expansive marketing campaign geared toward women, exploiting the germ anxieties connected to human touch, particularly the 'flies, fingers, and food' trio fundamental to the gospel of germs [27]. Cellophane was introduced as a disposable and transparent packaging film that was promoted as 'sanitary' [43]. Not only could the consumer see the product that they were buying, but they also found it *clean* and untouched by human hands. Figure 2.10 shows a 1932 advertisement claiming that Cellophane "beats Nature's best wraps", even better than the Australian Brown Onion. An excerpt from the caption below the image reads:

> It does the handsome thing by protecting many of the useful paper products. For example, the paper plates and cups that you can eat and drink from, the facial tissue that you use on your skin, and the note-paper you write on are wrapped in transparent Cellophane to keep them clean and sanitary, and to let you *see* what you buy.

DuPont initially targeted consumer products commonly associated with germs and uncleanliness, including meat, bread, and tobacco products, before the product became entirely ubiquitous [43]. Again, technology is used to achieve what nature couldn't.

In the 1956 advertisement shown in Figure 2.11, a young boy enveloped in cellophane is accompanied by the caption,

"Everything's at its best in Cellophane". Next to the child reads the note, "Mom says I'm so fresh and so clean (sometimes) – she ought to wrap me in Cellophane to keep me that way".

While the notion of encasing a child in plastic to preserve his cleanliness may seem absurd and irrational, this image reflects an emerging connection between transparency and perceptions of an ordered modernity, granting a sense of control over nature's capricious and unclean elements while enabling continuous human supervision and oversight. Shown in Figure 2.12, the concept seems a little more appropriate as a clean, transparent film, similar to Cellophane, delicately separates a child from the outdoors. This has become a tangible reality in our built environment of today, where glass is used to physically protect humans from nature while still allowing a perceived connection through the unobstructed view it affords. Through these visual representations, we come to understand the intricate interplay between transparency, modernity, and our desire for control in shaping indoor spaces. The transformative power of these ideas has permeated architectural design, erecting barriers between nature and the indoors while offering a nuanced perspective on our relationship with both.

Though we recognize the benefits of connecting to nature, the fear of the uncontrolled is still deeply embedded in Western culture. In the same way that cellophane played a transformative role in the food and grocery industry, glass became a trademark of modern architecture. In fact, Ian McCallum, the executive editor of *Architectural Review*, described the curtain wall as "the new vernacular" in 1957 [44]. Though the material is certainly not an invention of modernism, its use as a non-load bearing façade material emerged with reinforced concrete and steel framing, where the structural loads of the building could be pulled away from the façade. Additionally, mechanical systems absorbed the task of modulating the interior environment. As a result of this new generation of climate-independent buildings, Western cultural norms and thermal expectations spread around the world, driven by the forces of globalization. According to architectural historian Daniel Barber, "a conditioned, rather than designed, thermal interior was in this sense a signal characteristic of the Great Acceleration and of contemporary economic life" [45]. Ultimately, the universal adoption of air conditioning, sealed buildings, and curtain wall technology, all within the span of a single generation, has created incredible impacts on the urban skyline as well as on the human indoor experience.

Summary

Our relationship with the outdoors has shifted drastically throughout history, and cultural notions of health and hygiene have had a dramatic impact on how we shape the indoor environment. Despite significant improvements in urban infrastructure, the early 1900s brought a shift from public cleanliness to a focus on individual responsibility in public health. To create a society free of illness and disease, it was believed that mechanical conditioning could solve the evils of the world through technological evolution. By maximizing efficiency while also controlling the forces of nature, this notion laid the foundation for the current cultural epoch: the air-conditioning era. In describing this period of history, geologist Will Steffen stated that that these years "have without doubt seen the

Figure 2.11 Advertisement for DuPont Cellophane, "Everything's at its best in Cellophane", 1956. E.I. du Pont de Nemours & Company. Courtesy of Hagley Museum and Library.

most rapid transformation of the human relationship with the natural world in the history of humankind" [45]. This period formed tremendous social and behavioral habits that continue to drive our relationship with health, hygiene, nature, and buildings, representing the human tendency to conquer threats with technology. Similarly, the human connection to hygiene has been guided by multiple factors ranging from scientific reasoning and technological advancements to targeted marketing campaigns and cultural constructs.

The foundation for the air-conditioning era was built on three primary tenets: the development of a disease-free society, technical triumphalism, and the control of nature [11]. The next four chapters include contributions from a diverse group of experts on the interior environment, each with a unique perspective on future modes of human habitation, followed by a moderated conversation between the four authors.

Figure 2.12 Preschool-age child looking out the window, 2019. Photo by Gajus.

Note: An earlier version of this chapter was included in the proceedings for the ARCC Conference in Miami, entitled, "Modernity and Human Health: The Connection to Outdoor Air", co-authored with Traci Rose Rider.

References

[1] FordHeritage, dir. 2022. "Ford LTD Ad 1972." www.youtube.com/watch?v=9s6 FyT7Z3FQ.

[2] Saatchi & Saatchi. 2008. "Toyota – Badgers." *Ad Age*, February 3. https://adage.com/videos/toyota-badgers/524.

[3] Merchant, Carolyn. 2013. *Reinventing Eden: The Fate of Nature in Western Culture*. London: Taylor & Francis Group. http://ebookcentral.proquest.com/lib/uncc-ebooks/detail.action?docID=1154281.

[4] Appleton, Jay. 1975. *The Experience of Landscape*. London and New York: Wiley. https://catalog.lib.ncsu.edu/catalog/NCSU211097.

 [5] Mishra, A. K., M. G. L. C. Loomans, and J. L. M. Hensen. 2016. "Thermal Comfort of Heterogeneous and Dynamic Indoor Conditions – An Overview." *Building and Environment* 109 (November): 82–100. https://doi.org/10.1016/j.buildenv.2016.09.016.

 [6] Peng, You, Tao Feng, and Harry J. P. Timmermans. 2021. "Heterogeneity in Outdoor Comfort Assessment in Urban Public Spaces." *Science of The Total Environment* 790 (October): 147941. https://doi.org/10.1016/j.scitotenv.2021.147941.

 [7] Nikolopoulou, Marialena, and Koen Steemers. 2003. "Thermal Comfort and Psychological Adaptation as a Guide for Designing Urban Spaces." *Energy and Buildings, Special Issue on Urban Research* 35 (1): 95–101. https://doi.org/10.1016/S0378-7788(02)00084-1.

 [8] Ryan, Catherine O., William D. Browning, Joseph O. Clancy, Scott L. Andrews, and Namita B. Kallianpurkar. 2014. "Biophilic Design Patterns: Emerging Nature-Based Parameters for Health and Well-Being in the Built Environment." *ArchNet-IJAR: International Journal of Architectural Research* 8 (2): 62–75.

 [9] Parkinson, Thomas, and Richard de Dear. 2015. "Thermal Pleasure in Built Environments: Physiology of Alliesthesia." *Building Research & Information* 43 (3): 288–301. https://doi.org/10.1080/09613218.2015.989662.

[10] Moe, Kiel. 2007. "Compelling Yet Unreliable Theories of Sustainability." *Journal of Architectural Education* 60 (4): 24–30. https://doi.org/10.1111/j.1531-314X.2007.00105.x.

[11] Böer, Wulf. 2019. "Air-Conditioning, Architecture, and Modernism: On the History of the Controlled Environment, 1911–1952." Dissertation, ETH Zurich, Zurich, Switzerland. https://search.ebscohost.com/login.aspx?direct=true&db=ddu&AN=60C3EA6A585555E1&site=ehost-live.

[12] Cooper, Gail. 2002. *Air-Conditioning America: Engineers and the Controlled Environment, 1900–1960*. Baltimore: JHU Press.

[13] Bellamy, Edward. 1887. *Looking Backward: 2000–1887*.

[14] Ackermann, Marsha. 2002. *Cool Comfort: America's Romance with Air-Conditioning*. Washington, DC: Smithsonian Institution.

[15] Menzies, William Cameron, dir. 1936. "Things to Come." https://www.anthonyburgess.org/blog-posts/anthony-burgess-at-the-movies-things-to-come-dir-william-cameron-menzies-1936/.

[16] Thompson, Robert S. 2007. "The Air-Conditioning Capital of the World: Houston and Climate Control." In *Energy Metropolis: An Environmental History of Houston and the Gulf Coast*, edited by Martin V. Melosi and Joseph A. Pratt. Philadelphia: University of Pittsburgh Press. http://ebookcentral.proquest.com/lib/ncsu/detail.action?docID=2038827.

[17] Kalb, Bernard. 1955. "Air-Conditioned Life Brings Change." *The New York Times*, May 15.

[18] Bauckham, Richard. 2006. "Modern Domination of Nature." In *Environmental Stewardship*, 32–50. London: A&C Black.

[19] Brager, Gail S., and Richard J. de Dear. 2008. "Historical and Cultural Influences on Comfort Expectations." In *Buildings, Culture and Environment: Informing Local and Global Practices*, edited by Raymond J. Cole and Richard Lorch, 177–201. New York: John Wiley & Sons.

[20] Moe, Kiel. 2010. *Thermally Active Surfaces in Architecture*. Princeton: Princeton Architectural Press.

[21] Walker, Stanley. 1955. "Houston: Coolest Spot in U.S." *The New York Times*, May 15.

[22] Banham, Reyner. 1969. *The Architecture of the Well-Tempered Environment*. Chicago: University of Chicago.

[23] Kelley, Stephen J., and Dennis K. Johnson. 2013. "The Metal and Glass Curtain Wall: The History and Diagnostics." In *Modern Movement Heritage*. London: Taylor & Francis.

[24] Forbes, Peter. 2006. *The Gecko's Foot: Bio-Inspiration: Engineering New Materials from Nature*. Illustrated edition. New York: W. W. Norton & Company.

[25] Eldemery, I. M. 2009. "Globalization Challenges in Architecture." *Journal of Architectural and Planning Research* 26 (4): 343–54.

[26] Braham, William W. 2000. "Erasing the Face: Control and Shading in Post-colonial Architecture." *Interstices: Journal of Architecture and Related Arts*, 5(5), 104–113. https://doi.org/10.24135/ijara.v0i0.299.

[27] Tomes, Nancy. 1998. *The Gospel of Germs: Men, Women, and the Microbe in American Life*. Cambridge, MA: Harvard University Press.

[28] Blaser, Martin J. 2015. *Missing Microbes: How the Overuse of Antibiotics Is Fueling Our Modern Plagues*. First Picador edition. New York: Picador. https://catalog.lib.ncsu.edu/catalog/NCSU3511142.

[29] Karamanou, Marianna, George Panayiotakopoulos, Gregory Tsoucalas, Antonis A Kousoulis, and George Androutsos. 2012. "From Miasmas to Germs: A Historical Approach to Theories of Infectious Disease Transmission." *Le Infezioni in Medicina: Rivista Periodica di Eziologia, Epidemiologia, Diagnostica, Clinica e Terapia Delle Patologie Infettive* 5.

[30] Krieger, Nancy. 2011. *Epidemiology and the People's Health: Theory and Context*. Oxford: Oxford University Press.

[31] Damsky, Ellen. 2003. "A Way of Life: Saranac Lake and the 'Fresh Air' Cure for Tuberculosis." Ph.D., State University of New York, Binghamton. www.proquest.com/docview/305238703/abstract/7E0A9CDECAB74D42PQ/1.

[32] Douglas, Mary. 2002. *Purity and Danger: An Analysis of Concepts of Pollution and Taboo*. London: Routledge. https://doi.org/10.4324/9780203361832.

[33] Dancer, Stephanie J. 2013. "Infection Control in the Post-Antibiotic Era." *Healthcare Infection* 18 (2): 51–60. https://doi.org/10.1071/HI12042.

[34] Curtis, Valerie A. 2007. "Dirt, Disgust and Disease: A Natural History of Hygiene." *Journal of Epidemiology & Community Health* 61 (8): 660–4. https://doi.org/10.1136/jech.2007.062380.

[35] Pickering, Lucy, and Phillippa Wiseman. 2019. "Dirty Scholarship and Dirty Lives: Explorations in Bodies and Belonging." *The Sociological Review* 67 (4): 746–65. https://doi.org/10.1177/0038026119854244.

[36] Hoy, Suellen. 1996. *Chasing Dirt: The American Pursuit of Cleanliness*. New York: Oxford University Press. http://ebookcentral.proquest.com/lib/uncc-ebooks/detail.action?docID=3052065.

[37] Health and Cleanliness Council, dir. 1927. "Giro the Germ Episode 1." British Film Institute. https://player.bfi.org.uk/free/film/watch-giro-the-germ-episode-1-1927-online.

[38] Boots Company Archive, dir. 1935. "See How They Won." Boots Company Archive, Nottingham, UK.

[39] Stark, James F., and Catherine Stones. 2019. "Constructing Representations of Germs in the Twentieth Century." *Cultural and Social History* 16 (3): 287–314. https://doi.org/10.1080/14780038.2019.1585314.

[40] Tomes, Nancy. 2000. "The Making of a Germ Panic, Then and Now." *American Journal of Public Health* 90 (2): 191–8. https://doi.org/10.2105/ajph.90.2.191.

[41] Vinikas, Vincent. 1989. "Lustrum of the Cleanliness Institute, 1927–1932." *Journal of Social History* 22 (4): 613–30.

[42] Wohl, Anthony S. 1983. *Endangered Lives: Public Health in Victorian Britain.* Cambridge, MA: Harvard University Press.

[43] Hisano, Ai. 2017. "Cellophane, the New Visuality, and the Creation of Self-Service Food Retailing." *SSRN Electronic Journal* 17–106. https://doi.org/10.2139/ssrn.2973544.

[44] Yeomans, David. 1998. "The Pre-History of the Curtain Wall." *Construction History* 14: 59–82.

[45] Barber, Daniel A. 2020. *Modern Architecture and Climate: Design Before Air Conditioning.* Princeton: Princeton University Press.

3 The Microscopic World of Building Science

Sarah Haines

Microorganisms, also known as fungi (mold and yeast) and bacteria, are everywhere. They are found outside, in soil, on plants, in the food we eat, in animals, in and on our bodies, on surfaces, and in our homes. Microorganisms, or *microbes* for short, are also found in the most extreme environments on our planet, such as hot geothermal vents as well as in frigid high-pressure water at the bottom of the ocean where it seems like nothing should survive. While this may seem frightening (they're everywhere!), microorganisms serve numerous beneficial purposes in our everyday lives. The "human microbiome", the complex collective of the 39 trillion microorganisms in and on our bodies, shape and protect our health and well-being [1]. Microorganisms in our gut have recently gained attention for their important connection to human health, as they aid in key metabolic functions, immune system education, and protection against pathogenic bacteria. In fact, the gut-brain axis, which refers to the connection between bacteria in our gut and our brain, maintains critical functions of the gastrointestinal and central nervous system [2]. Additionally, microbes are critical to life on Earth as they perform fundamental functions to maintain soil health such as nutrient cycling, crop residue degradation, and plant growth stimulation. Another important function of microorganisms is in cheese and bread making. The distinctiveness of a cheese is distinguished by the types of microorganisms that make the cheese, such that certain species of bacteria are used to make Swiss cheese while others are used to make cheddar. Yeast, a type of fungus, is critical to bread making as the yeasts feed on sugar, releasing carbon dioxide which causes bread to rise. Without these microorganisms, we would forever be deprived of grilled cheese!

We often think of microorganisms as harmful contaminants, something to remove from our homes and spaces using anti-microbial soaps and cleaners. Particularly with the COVID-19 pandemic, there has been a surge to rid our surfaces of viruses and bacteria. However, it is important to examine the balance of harmful microorganisms and those that may have beneficial and protective effects.

DOI: 10.1201/9781003398714-4

Shaping an Indoor Microbiome

Microorganisms are everywhere, including in our indoor built environment where we spend roughly 90% of our time. The concept of an indoor built environment is relatively new, as we are the only known organisms to modify our habitats with complex systems and diverse structures. The term "built environment" has origins in anthropology and social science literature starting in the 1970s [3], around the same time that central air conditioning usage in housing became more commonplace in the United States. In our built environment, we are consistently exposed to a variety of chemical pollutants, particulate matter, and microorganisms. Just as we have a human microbiome, there exists an indoor microbiome [4]. The indoor microbiome is a complex collective community of viruses and living microorganisms as well as their fragments and byproducts, such as microbial volatile organic compounds (mVOCs) or mycotoxins. Sources of indoor microorganisms include humans, plants, pets, plumbing, HVAC, and the outdoor environment [4]. People and the outdoor environment particularly have a critical impact on shaping our indoor microbiome.

Humans Influence Indoor Spaces

Humans are a major contributor to the indoor microbiome, as we harbor roughly one trillion microorganisms on our skin [5] and release roughly 37 million bacterial genome copies and 7.3 million fungal genome copies per person-hour [6]. This means that we are constantly shedding microorganisms and populating the indoor microbiome. This contribution not only impacts the *quantity* of microorganisms indoors, but also the *types* of microorganisms, as the bacteria found in indoor environments are commonly associated with human-associated bacteria [7]. As such, when humans move into new spaces, they have been found to populate these environments with their own human-associated microorganisms and biological signatures within three hours of habitation. Through "The Home Microbiome Project", researchers followed seven diverse US families and their homes for a period of six weeks. Three of the families moved to a new home during this study, and samples were taken immediately before and after moving. In less than one day, the microbiome of the new house consisted of the *same* microbial signatures as the old house. When compared across families and homes, each home had a distinct indoor microbiome. So distinct that researchers were able to accurately match people to their correct home based on microbiome alone [8]. Similar results have been found in additional studies where the presence of humans had the greatest influence over the types of microbes present [9,10]. Though many other sources influence indoor microbiomes, which will be discussed here, the presence of humans tend to be a dominating factor. Though we are decades away from active usage, these distinct microbial signatures may be utilized in forensic investigations to identify the presence of an individual at the scene of a crime.

Not only does the human microbiome populate and dictate common indoor microorganisms, but human architectural design also influences indoor microbial communities. As defined in previous work, function (activities and practices that a building and its spaces serve), form (geometry of a building and the spaces within it), and organization (the spatial relationships among indoor spaces) each dictate the overall indoor microbiome [11]. Researchers determined that both the ventilation source (dispersal of outside organisms indoors) and the conditions within the building (temperature, relative humidity, floor type, and space type) contributed to the indoor microbiome [12,13].

Additionally, different materials used in construction are likely to influence microbial communities, particularly at increased moisture levels. In a comparison between drywall and carpet exposed to elevated (over 80% relative humidity) moisture conditions, the quantity of fungal (mold) growth differed at varying humidity levels (Figure 3.1) [14]. Fungal growth in carpet is found to occur over 75% relative humidity (RH), while growth does not occur until over 85% RH in drywall [14]. Additionally, different types of carpet material (nylon, olefin, and wool) resulted in varying levels of fungal growth at elevated humidity conditions, with the most growth found on nylon, and the least on olefin [15]. Recent work has highlighted that similar bacterial communities are found on building materials within a home, though bacterial communities will differ depending on their location within a room. For example, a floor will contain a much more diverse community composition when compared to that of a ceiling [16]. Similarly,

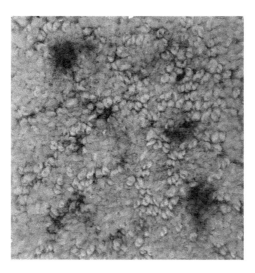

Figure 3.1 Active mold growth on carpet exposed to over 85% relative humidity for 14 days. Photo by author.

the microbial communities under the kitchen sink may differ from the communities in the bathroom by the toilet or in a bedroom by the pillow. Other work has determined differences in fungal community composition on varying materials (carpet and drywall) [14]; however, these differences may be due to the moisture content of the materials themselves or the location of collection. Within these studies, moisture availability is a critical factor to growth. Therefore, a major concern when choosing building materials is likely the availability for water uptake and the likelihood of moisture in the building material location, as increasing access to damp conditions may change the microbial profile of certain indoor materials such as wood, drywall, and carpet for the worse. Many companies offer *anti-microbial* or *probiotic* materials; however, more research is required in this area. Bamboo has been lauded as a potentially beneficial building material, as it is quick to grow and easy to modify with graphene/zinc oxide coatings, nanoscale metal adhesion, and thermal treatments to protect from harmful mold growth [17,18]. This may be particularly useful in hot, wet environments impacted by climate change that need sustainable and efficient means to construct healthy homes. The link between architectural design and indoor microbiome is limited, as the mechanisms that shape these spaces are still not well defined. Continued research and collaborations between architects, engineers, and microbiologists would aid in elucidating these challenges.

How the Outdoor Environment Influences Our Indoor Environment

Though people tend to be a major contributor of bacterial communities indoors, the outdoor environment is a major contributor of fungal communities (as well as some bacterial communities) in indoor environments. Indoor fungal organisms are commonly associated with outdoor sources and seasonal influences [19]. There are typically increased indoor fungal concentrations in the fall and summer, which are likely related to the increase of outdoor fungal concentrations that occur due to variations in humidity, temperature, growing plants, and other outdoor activities that occur in these seasons. Common indoor fungal organisms may release potentially harmful microbial volatile organic compounds (mVOCs), mycotoxins, and allergens.

Geographic location also influences indoor microbial communities (both fungi and bacteria) due to differences in atmosphere, land type, and climate. Due to these distinct differences, researchers have been able to accurately determine in which city a certain microbiome sample originated based on the microbial profile [16]. Shelton (2002) found regional variability in terms of indoor microbial profiles across the US. The Southwest (Arizona, New Mexico, Oklahoma, and Texas), Far West (California, Nevada, Oregon, and Washington), and Southeast (Alabama, Arkansas, Georgia, Florida, Kentucky, Louisiana, Mississippi, North Carolina, South Carolina, Tennessee, and Virginia) United States were determined to be home to the most abundant indoor microbial profiles [20]. In France, similar work has been conducted to analyze indoor microbial profiles in which composition of indoor microbiomes were dependent on landscape with certain species found in cold-humid landscapes and others in more open, hot-dry landscapes. Therefore, when designing indoor spaces, we must be cognizant of the geographic location and how a home environment may be modified due to the presence of certain microorganisms.

The microbiome of a damp building is often distinct from that of a non-damp building, with moisture being the critical limiting factor to fungal growth. Recent work using DNA sequence-based approaches was able to classify the mold status of a building based on microbial composition [21], though current methods of mold detection, prediction, and prevention are still underway. As temperature and humidity increase due to climate change, not only will our outside environment be impacted, but our indoor environment will be as well. Researchers of indoor air quality are already postulating the impacts of how our indoor microbiomes may transform with a changing climate. Increased incidence of hurricanes and extreme precipitation may cause damaged building materials and indoor flooding, which will likely lead to a surge in fungal growth. Additionally, indoor microbiome profiles of areas that are historically cool and dry may change and become more similar to microbial profiles historically found in hot and wet regions. This will likely cause changes in allergies as well as increased potential for indoor fungal growth. In a recent climate study, scientists predict that Philadelphia, Pennsylvania may share a similar climate to that currently found in Memphis, Tennessee by 2080 [22]. As such, it is likely that microbial communities normally associated with Memphis may become more prevalent in Philadelphia. Research in this area is currently underdeveloped and it will be important to examine how climate change shapes our indoor microbiomes, which in turn will impact human health.

Are Probiotics the Future of the Indoor Microbiome?

Our indoor microbiomes are shaped by a variety of factors, with certain microbial exposures leading to potentially negative health impacts. Recently the "hygiene hypothesis", which states that overly clean environments fail to provide necessary exposures to microorganisms to *educate* our immune systems so that our bodies can respond to infectious organisms, has taken a stronger hold in discussions about our indoor microbiome. Exposure to a diverse set of microorganisms at a young age has been associated with decreased asthma risk [23]; however, it is not clear if only diversity is needed or whether a specific microbial community made up of a beneficial mixture is required. Recent work has suggested that exposure to farm-associated microorganisms in indoor spaces, for example, can have positive effects. A study of Amish and Hutterite communities in the United States found that the prevalence of asthma and allergic sensitization was 4–6 times lower in Amish children than in Hutterite children [24]. Though both Amish and Hutterite communities have similar farming practices, Hutterite families live away from their farms, while Amish families live directly within their farm property and are therefore likely exposed to common farm microbiota. An additional study was conducted in both a Finnish birth cohort and German cohort, in which incidence of asthma from children on farm and non-farm homes were compared. In these studies, children who were exposed to farm-like micro-biota had decreased asthma risk [25]. These exposures are considered protective, meaning that exposure to certain microorganisms at an early age may have shielding effects against disease development.

Other studies have also analyzed potential protective effects of microorganisms, such as a dog-associated mycobiome (fungal-specific microbiome) in protecting against asthma and allergic airway diseases [26]. Exposure to dog-associated microorganisms early in life has led to protective effects against allergic disease development [27,28]. Additionally, it is suggested that dogs, such as the one shown in Figure 3.2, disturb outside soil and may carry additional protective microorgan-isms inside our built environment. To further examine how potential outdoor microbiota may have protective or "beneficial" effects on indoor environments, a recent study examined the helpfulness of including plants indoors. Introductions of indoor plants were found to increase microbial diversity [29,30]; however,

Figure 3.2 Soil associated microbes brought into the home by dogs can have beneficial effects. Photo by Elizabeth L. McCormick.

it is still not clear whether this is considered beneficial to human health. Plants have long been thought of as a clear fix to improving indoor air quality, and though plants may provide some psychological benefits and contribute to some air cleaning capabilities, this research is still unclear, particularly how the plant microbiome impacts indoor spaces.

Though many aspects of our indoor microbiome remain unclear, one thing is certain: the future development of the built environment must account for the indoor microbiome when considering the design and operation of a building. Continued science is needed to determine which mixtures of indoor microorganisms may be beneficial in promoting a healthy lifestyle and protective of certain other harmful microorganisms and diseases. In the future, once more connections between the protective effects of certain microbial exposures are identified, it may be possible to "seed" indoor microbiomes with probiotics (beneficial microorganisms) to enhance overall human health. One vision is seeding materials in indoor spaces with these beneficial organisms and ideally positively changing the overall microbiome. However, many of the protective effects only happen in the first months of life, so this work would likely be critical for prenatal, infants, and children.

Equity and Regulation

While microorganisms are ubiquitous in our indoor environments, certain exposures may be harmful to human health. Roughly 25 million Americans suffer from asthma; of these, 4.6 million cases are associated with exposure to mold (fungi) and dampness indoors. This exposure, which is particularly common in low socioeconomic status communities (low-SES), as well as Indigenous communities throughout North America [31–34], is associated with stuffy nose, wheezing, itchy eyes, and upper respiratory diseases. Housing in low-SES communities has long been associated with increased risk of health concerns; however, indoor environmental quality is not often discussed in an environmental justice context. Social determinants of health such as poor housing quality, older building age, and poor neighborhood quality are associated with and more prevalent in non-white and low-SES communities. As a result, these communities are at greater health risk for incidences of asthma, respiratory disease, and cancer. In the United States there is a high prevalence of asthma and asthma morbidity in children in low-SES neighborhoods, particularly in urban cities, as household allergen exposure is influenced by socioeconomic status [35]. Additionally, when compared to white children, non-Hispanic black and Puerto Rican children are more likely to have asthma [36]. Ensuring proper housing and indoor environmental quality is critical to maintaining proper human health.

Though much of our exposure to indoor microorganisms occurs through resuspension of dust and subsequent inhalation, there are generally no accepted threshold limits or defined national standards that outline the allowable quantity of bacteria, fungi, or virus in indoor air. With the COVID-19 pandemic there was more focus on improvements in ventilation and movement of air through buildings, such as strengthening recommendations for the number of air changes per hour (i.e., the number of times the total volume of air in a room is removed and replaced each hour) through a space. The American Society of Heating, Refrigerating and Air-Conditioning Engineers (ASHRAE) advises that a home should receive 0.35 air changes per hour as the minimum ventilation rate; however, this value changes depending on the space, with restaurants and sport facilities recommended to have at least 6–8 air changes per hour due to higher occupancy capacity [37]. However, there is not a quantifiable value to safely state the maximum recommended allowable concentration of microorganisms in the air that exists for other air pollutants, such as carbon monoxide, radon, formaldehyde, and volatile organic compounds in certain household products regulated under the Clean Air Act, a US federal law that regulates sources of air emissions [38]. As we continue to improve knowledge of the microorganisms that shape our indoor spaces, recommendations for maximum exposure levels will improve overall harmful exposures and prevent health effects.

How we use our buildings also contributes to the overall microbiome. It is important that buildings are designed for ease, but also that we are promoting beneficial microbial interactions and limiting harmful mold growth due to increased moisture. As the climate changes throughout the globe, our indoor environments are also likely to change. Accounting for these changes while also utilizing beneficial microorganisms in indoor environments may be the key to developing healthy, sustainable, and equitable indoor spaces. Though much more work is needed to determine the right balance of too much and too little indoor microbial exposures, there is clear evidence that certain microbial exposures have beneficial health implications. For the past century, special efforts have been made to ensure that buildings were void of microorganisms and sealed tight to keep out contaminants and prevent air leakage. Despite the best intentions, this is likely influencing our indoor environmental quality for the worse. A balance of beneficial microorganisms (potentially those from outdoor sources) in indoor spaces may positively influence occupant health and well-being. Additionally, by tightening and weatherproofing a house we may be indirectly increasing buildup of moisture and mold growth, influencing exposure of harmful microorganisms. It is therefore this balance between promoting beneficial microbes and reducing exposure to harmful molds and allergens that will pave the future for indoor environmental quality.

References

[1] Sender R, Fuchs S, Milo R. Revised estimates for the number of human and bacteria cells in the body. *PLOS Biology*. 2016;14(8):e1002533. doi:10.1371/JOURNAL.PBIO.1002533

[2] Morais LH, Schreiber HL, Mazmanian SK. The gut microbiota–brain axis in behaviour and brain disorders. *Nature Reviews Microbiology*. 2020;19(4):241–255. doi:10.1038/s41579-020-00460-0

[3] Moffatt S, Kohler N. Conceptualizing the built environment as a social–ecological system. *Building Research and Information*. 2008;36(3):248–268. doi:10.1080/09613210801928131

[4] Prussin AJ, Marr LC. Sources of airborne microorganisms in the built environment. *Microbiome*. 2015;3:78. doi:10.1186/s40168-015-0144-z

[5] Luckey TD. Introduction to intestinal microecology. *The American Journal of Clinical Nutrition*. 1972;25(12):1292–1294. doi:10.1093/AJCN/25.12.1292

[6] Qian J, Hospodsky D, Yamamoto N, Nazaroff WW, Peccia J. Size-resolved emission rates of airborne bacteria and fungi in an occupied classroom. *Indoor Air*. 2012;22(4):339–351. doi:10.1111/j.1600-0668.2012.00769.x

[7] Hospodsky D, Qian J, Nazaroff WW, et al. Human occupancy as a source of indoor airborne bacteria. *PLOS One*. 2012;7(4):34867. doi:10.1371/JOURNAL.PONE.0034867

[8] Lax S, Smith DP, Hampton-Marcell J, et al. Longitudinal analysis of microbial interaction between humans and the indoor environment. *Science*. 2014;345(6200):1048–1052. doi:10.1126/SCIENCE.1254529

[9] Haines SR, Siegel JA, Dannemiller KC. Modeling microbial growth in carpet dust exposed to diurnal variations in relative humidity using the "time-of-wetness" framework. *Indoor Air*. 2020;30(5):978–992. doi:10.1111/ina.12686

[10] Täubel M, Rintala H, Pitkäranta M, et al. The occupant as a source of house dust bacteria. *Journal of Allergy and Clinical Immunology*. 2009;124(4):834–840.e47. doi:10.1016/J.JACI.2009.07.045

[11] Kembel SW, Meadow JF, O'Connor TK, et al. Architectural design drives the biogeography of indoor bacterial communities. *PLOS One*. 2014;9(1). doi:10.1371/journal.pone.0087093

[12] Kembel SW, Meadow JF, O'Connor TK, et al. Architectural design drives the biogeography of indoor bacterial communities. *PLOS One*. 2014;9(1):e87093. doi:10.1371/JOURNAL.PONE.0087093

[13] Meadow JF, Altrichter AE, Kembel SW, et al. Indoor airborne bacterial communities are influenced by ventilation, occupancy, and outdoor air source. *Indoor Air*. 2014;24(1):41–48. doi:10.1111/ina.12047

[14] Haines SR, Hall EC, Marciniak K, et al. Microbial growth and volatile organic compound (VOC) emissions from carpet and drywall under elevated relative humidity conditions. *Microbiome*. 2021;9. Published Online.

[15] Nastasi N, Haines SR, Xu L, et al. Morphology and quantification of fungal growth in residential dust and carpets. *Building and Environment*. 2020;174:106774. doi:10.1016/j.buildenv.2020.106774

[16] Chase J, Fouquier J, Zare M, et al. Geography and location are the primary drivers of office microbiome composition. *mSystems*. 2016;1(2). doi:10.1128/msystems.00022-16

[17] Nurdiah EA. The potential of bamboo as building material in organic shaped buildings. *Procedia – Social and Behavioral Sciences*. 2016;216. Published Online. doi:10.1016/j.sbspro.2015.12.004

[18] Zhang J, Zhang B, Chen X, et al. Antimicrobial bamboo materials functionalized with ZnO and graphene oxide nanocomposites. *Materials*. 2017;10(3). doi:10.3390/ma10030239

[19] Adams RI, Miletto M, Taylor JW, Bruns TD. Dispersal in microbes: Fungi in indoor air are dominated by outdoor air and show dispersal limitation at short distances. *The ISME Journal*. 2013;7(7):1262–1273. doi:10.1038/ismej.2013.28

[20] Shelton BG, Kirkland KH, Flanders WD, Morris GK. Profiles of airborne fungi in buildings and outdoor environments in the United States. *Applied and Environmental Microbiology*. 2002;68(4):1743–1753. doi:10.1128/AEM.68.4.1743-1753.2002/

ASSET/C54805F5-458B-4BAE-9845-A81C19CD26C0/ASSETS/GRAPHIC/
AM0421595003.JPEG

[21] Hegarty B, Pan A, Haverinen-Shaughnessy U, Shaughnessy R, Peccia J. DNA Sequence-based approach for classifying the mold status of buildings. *Environmental Science & Technology*. 2020;54(24):15968–15975. doi:10.1021/ACS.EST.0C03904

[22] Fitzpatrick MC, Dunn RR. Contemporary climatic analogs for 540 North American urban areas in the late 21st century. *Nature Communications*. 2019;10(1):1–7. doi:10.1038/s41467-019-08540-3

[23] Dannemiller KC, Mendell MJ, Macher JM, et al. Next-generation DNA sequencing reveals that low fungal diversity in house dust is associated with childhood asthma development. *Indoor Air*. 2014;24(3):236–247. doi:10.1111/ina.12072

[24] Stein MM, Hrusch CL, Gozdz J, et al. Innate immunity and asthma risk in Amish and Hutterite farm children. *New England Journal of Medicine*. 2016;375(5):411–421. doi:10.1056/NEJMOA1508749/SUPPL_FILE/NEJMOA1508749_DISCLOSURES. PDF

[25] Kirjavainen PV, Karvonen AM, Adams RI, et al. Farm-like indoor microbiota in non-farm homes protects children from asthma development. *Nature Medicine*. 2019;25(7):1089–1095. doi:10.1038/s41591-019-0469-4

[26] Rush RE, Dannemiller KC, Cochran SJ, et al. Vishniacozyma victoriae (syn. Cryptococcus victoriae) in the homes of asthmatic and non-asthmatic children in New York City. *Journal of Exposure Science & Environmental Epidemiology*. 2021;32(1):48–59. doi:10.1038/s41370-021-00342-4

[27] Fujimura KE, Johnson CC, Ownby DR, et al. Man's best friend? The effect of pet ownership on house dust microbial communities. *The Journal of Allergy and Clinical Immunology*. 2010;126(2):410. doi:10.1016/J.JACI.2010.05.042

[28] Campo P, Kalra HK, Levin L, et al. Influence of dog ownership and high endotoxin on wheezing and atopy during infancy. *The Journal of Allergy and Clinical Immunology*. 2006;118(6):1271. doi:10.1016/J.JACI.2006.08.008

[29] Berg G, Mahnert A, Moissl-Eichinger C. Beneficial effects of plant-associated microbes on indoor microbiomes and human health? *Frontiers in Microbiology*. 2014;5. Published Online. doi:10.3389/fmicb.2014.00015

[30] Dockx Y, Täubel M, Bijnens EM, et al. Indoor green can modify the indoor dust microbial communities. *Indoor Air*. 2022;32(3):e13011. doi:10.1111/INA.13011

[31] Jones R, Recer GM, Hwang SA, Lin S. Association between indoor mold and asthma among children in Buffalo, New York. *Indoor Air*. 2011;21(2):156–164. doi:10.1111/j.1600-0668.2010.00692.x

[32] Kovesi T, Mallach G, Schreiber Y, et al. Housing conditions and respiratory morbidity in Indigenous children in remote communities in Northwestern Ontario, Canada. *Canadian Medical Association Journal*. 2022;194(3):E80–E88. doi:10.1503/CMAJ.202465

[33] Bryant-Stephens T. Asthma disparities in urban environments. *Journal of Allergy and Clinical Immunology*. 2009;123(6):1199–1206. doi:10.1016/j.jaci.2009.04.030

[34] Thacher JD, Gruzieva O, Pershagen G, et al. Mold and dampness exposure and allergic outcomes from birth to adolescence: data from the BAMSE cohort. *Allergy: European Journal of Allergy and Clinical Immunology*. 2017;72(6):967–974. doi:10.1111/all.13102

[35] Camacho-Rivera M, Kawachi I, Bennett GG, Subramanian SV. Associations of neighborhood concentrated poverty, neighborhood racial/ethnic composition, and indoor allergen exposures: A cross-sectional analysis of Los Angeles households, 2006–2008. *Journal of Urban Health.* 2014;91(4):661–676. doi:10.1007/S11524-014-9872-9/TABLES/4

[36] Bryant-Stephens T. Asthma disparities in urban environments. *Journal of Allergy and Clinical Immunology.* 2009;123(6):1199–1206. doi:10.1016/J.JACI.2009.04.030

[37] Standards 62.1 & 62.2. Accessed June 13, 2022. www.ashrae.org/technical-resources/bookstore/standards-62-1-62-2

[38] Indoor Air Quality (IAQ) | US EPA. Accessed June 13, 2022. www.epa.gov/indoor-air-quality-iaq

4 The Building/Occupant Organism

Marcel Harmon

Assessing the built environment's impact on health and productivity is a complex undertaking. The built environment is a multifaceted entity, and it can be useful to think of individual facilities as building/occupant organisms. Such organisms are composed of individual occupants, each person with associated physiological and psychological needs in varying degrees of alignment with the needs of other occupants, as shown in Figure 4.1. There are also the organizations housed within the facility, made up of individual occupants, and subject to their own policies and conventions, in varying degrees of alignment with the cultural norms of the larger encompassing communities and society to which they belong. In addition to the individuals and groups (organizations) are the environments that shape, constrain, facilitate, or hamper the desired actions of the individuals and organizations occupying them. This environment is both the physical building itself and the social/cultural environment, which consists of the various organizational cultural norms present, as well as those of the larger encompassing communities and society. Together these are what make up the building/occupant organism.

But the design, construction, and operation of this organism occurs within a larger nested hierarchy of groups up to the global level, each with varying needs and goals and in shifting degrees of alignment and conflict. The existence and functioning of the building/occupant organism within the context of these nested hierarchies impacts its performance, further impacting the health, wellness, and productivity of all involved. The vast majority of our evolutionary history has been spent in small groups without nested hierarchies and in limited contact with people outside of their group, and in physical environments vastly different from our contemporary indoor environments, i.e., predominantly outdoors [1–3]. It is only within the last several thousand years that we began to congregate in urban environments and only within the last 100+ years that we've begun to spend most of our time indoors within artificially controlled environments, heated, cooled, and ventilated by burning fossil fuels. This too shapes and constrains what the optimal configuration of the building/occupant organism should be. This chapter will discuss why it's important to take our evolutionary history into account

DOI: 10.1201/9781003398714-5

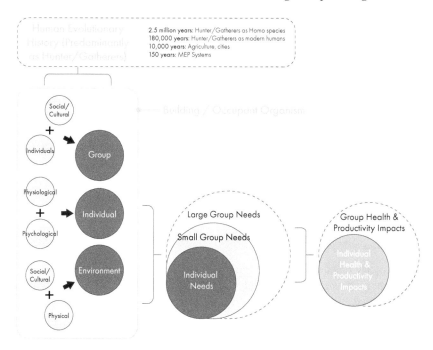

Figure 4.1 The complexities associated with the impacts of the built environment on health and productivity. Image by author.

when designing the building/occupant organism, focusing on the concept of evolutionary mismatches, and lay out a process for doing so. Put very simply, the insides of our contemporary environments should likely better mimic the outside of our evolutionary past.

Health and Productivity Are Proxies for Evolutionary Fitness

From an evolutionary perspective, built environments that inadequately account for our evolutionary history may also negatively impact our evolutionary fitness – our *success* reflective of how well we're adapted to the environment(s) we occupy [1,4]. And we can use the impacts on our physical and mental health, as well as our productivity, as a proxy for the impacts on our evolutionary fitness [5]. However, the complexity of the building/occupant organism often results in conflicts that occur among the various aspects of building, occupant, and organizational performance as well as between the different levels of the larger nested hierarchy of groups from Figure 4.1. Traits (biological and cultural) or environments (physical and social) that are adaptive for individuals may negatively impact the fitness of the organizations they're part of. Similarly, traits or environments that positively impact one aspect of individual health and productivity may negatively impact another aspect [6–11].

For example, some research has indicated that increasing ventilation levels (assuming good quality outdoor air is present) can a) increase individual cognitive function through a decrease in levels of carbon dioxide (CO_2) and particulate matter ($PM_{2.5}$) [12,13] and b) increase individual health through a decrease in the concentration levels of pathogens, pollutants, and toxins [14–16]. But if done poorly (relative to thermal comfort and/or energy performance) this can also a) decrease individual performance and health through increased thermal discomfort [17–20], b) decrease the individual's organization's fitness via increased utility costs and/or work orders [16,21], and c) decrease individual and organization productivity and health by increasing conflicts over thermal comfort and indoor air quality (IAQ) control [22].

Evolutionary Mismatches

Evolutionary mismatches are occurrences where an organism's traits that were adaptable in the *ancestral* environment(s) they originally evolved within end up being less adaptive or even *dysfunctional* in a different environment [23–25]. For example, the human circadian system (a system of traits) adapted to optimally function using cues (patterns of light and dark) from the natural 24-hour day/night cycle. However, research has shown that contemporary environments, which are often brightly illuminated well after sunset as we stare into the screens of our electronic devices late into the evening, negatively impact the functioning of our circadian system and ultimately our health [26–28]. It is mismatched to aspects of our current environments and lifestyles, and this is particularly problematic for shift workers.

While the basic concept is fairly simple, real-world assessments can be quite complex considering the three types of human factors – physiological, psychological, and social/cultural – and how we have been shaped by our evolutionary history. In addition, while a "new" environment may be aligned with one set of an organism's or group's traits (i.e., ASHRAE 62.1 ventilation minimums generally aligned with human olfactory-related traits), another set of traits from the same organism or group may be poorly adapted to that same "new" environment (i.e., ASHRAE 62.1 ventilation minimums potentially not aligned with human cognition related traits under certain contexts). As a result, it can get complicated. Considering humans have spent the vast majority of our evolutionary history outdoors, it isn't surprising that aspects of modern society, including our built environments where we spend over 90% of our time [29], might be dysfunctional and negatively impacting aspects of our health and productivity. That certainly includes the technology we incorporate within our built environments (such as the blue light-emitting screens of our electronic devices mentioned earlier). To dig into this in more detail, we'll evaluate ionization technology as an example in the next section.

Ionization Technology as an Evolutionary Mismatch

While our evolutionary history would suggest that finding more ways to bring the outdoors indoors should be beneficial, such as through increased ventilation and the use of operable windows, our modern world's poor outdoor air quality in many locations and at different times of the year (e.g., wildfire season) can negatively impact human health. Cleaning the air becomes more important in these situations and air filtration is a proven strategy. Another option often proposed are emergent additive air cleaning technologies (also referred to as electronic air cleaning technologies). A wide array of electrically powered air-cleaners exist that use varying types of indoor air chemistry to remove particles from airstreams or to inactivate pathogens [30]. Overall, this broad category of technology lacks adequate industry testing standards and regulations (as well as adequate supporting peer-reviewed research) [30,31] often has limited performance effectiveness in real-world conditions [30,32–35], and is potentially harmful in some contexts due to the air chemistry created by the tech's varying primary and secondary emissions (byproducts) [36–41].

One common category of electronic air cleaners is ionization technology. Such equipment generally produces negative ions only or both positive and negative ions, either within a space or at an air handling unit where they are then delivered to a space through the mechanical system's ductwork via the airstream. The ions react with water vapor and oxygen to create free radicals, or Reactive Oxygen Species (ROS). Once these ROS are created, the intent is for them to kill (or inactivate) pathogens in the air and on surfaces, decompose volatile organic compounds (VOCs), and by causing particles to cluster or agglomerate into larger particles, both increase the effectiveness of filter media and increase the rate at which particles drop out of the air. There is some evidence of ionization having limited success accomplishing these intents in a narrow range of environmental conditions for some makes and models of technology [32–34]. The only intended byproducts of this indoor air chemistry are benign water and carbon dioxide molecules resulting from the decomposition of VOCs. In reality, the ROS can interact with multiple other compounds in the indoor air until there aren't any ROS left. As a result, harmful byproducts can be created, such as ozone (O_3), nitrogen oxides (NOx), and various VOC oxidation intermediates, even if the targeted particle and compound concentration levels happen to be reduced (which isn't always the case) [32–34,36,40,41]. Setting aside the potential negative health impacts from exposure to the ROS and harmful byproducts, another legitimate question to ask is what impact does the exposure to these ions themselves have on human health?

This is a complex question involving such things as frequency of exposure, amount of exposure, age, pre-existing conditions, and environmental conditions. One way to categorize such complexities for further analysis is to use Biologist Nikolaas Tinbergen's four characterizations of traits [23,25,42,43]. These consist of the following:

1. Function: The reason(s) a trait or system of traits exists, typically common to a species. Successfully fulfilling the function increases the fitness level, or general health/wellness, of the individual or group in question.
2. Mechanisms: The intricate details of how the traits or systems of traits specifically work. The nature of the mechanisms is typically driven by the environmental conditions in which the population of individuals evolved.
3. Development: How the traits or systems of traits develop and change over the course of an individual's lifetime (also determined by our evolutionary history). They're often not fixed, as physiological systems tend to function less efficiently with age. The systems of the very young also tend to be more vulnerable to changes in the environment that impact how efficiently the mechanisms operate.
4. Heritable history: How traits or systems of traits evolve over the history of our species, and our species' ancestors, and how that history compares to the evolution of similar systems in other species. At various points within that history, the continued development of these traits was constrained by what came before.

Breaking traits, or systems of traits, down into these four characterizations can paint a clearer picture of how our evolutionary history, including past environments, shaped them to optimally function and maximize our fitness. It can provide greater insight into how various technologically influenced environmental conditions could negatively impact that functioning, our health, and ultimately our fitness. One of our physiological systems potentially impacted by direct exposure to positive or negative ions produced by ionization technology is the cardiovascular autonomic nervous system. Applying these four characterizations to this system results in the following:

Function

The ultimate reason for the system's evolution is to regulate heart rate and blood pressure in the short term to cope with everyday situations (such as rushing to meet a deadline or scrambling to get away from a predator) [44].

Mechanisms

The mechanistic basis of the system in humans consists of two branches of the autonomic nervous system: a) the sympathetic nervous system (SNS) which releases hormones (catecholamines – epinephrine and norepinephrine) to accelerate heart rate and b) the parasympathetic nervous system (PNS) which releases the hormone acetylcholine to slow heart rate [44]. We would want to know if the proximate conditions of ion exposure created by this technology negatively impact the functioning of these mechanisms, to the point that health and fitness are negatively impacted.

Development

The development of the cardiovascular autonomic nervous system over the course of an organism's lifetime may indicate that it operates differently at the various stages of human development or is impacted differently by the environment. In addition to impacts on the autonomic nervous system itself, looking at the cardiorespiratory system more broadly, the gas exchange alveolar surface area of the human lung increases up to 20-fold by the time one is an adult, out of proportion with the concurrent increase in chest size [45]. But some data suggests exposure to certain environmental factors, like ozone, may stunt this growth (though growth may recover depending on the stage of development and length of exposure) [45]. And the efficiency of gas exchange across the alveolar membrane can decrease with age, exacerbated by certain environmental exposures over one's lifetime [46] (potentially such as exposure to ions above a certain concentration level for a certain amount of time).

Heritable History

And finally, its heritable history refers to how the cardiovascular autonomic nervous system evolved over the history of our species and our species' ancestors, and how it compares to the evolution of similar systems in other species. For example, again looking at the cardiorespiratory system more broadly, a contributing factor to the decreasing efficiency of gas exchange across the alveolar membrane with age is likely the mammalian lung's combining gas exchange and ventilatory (physical breathing) functions in the same lung tissue [47]. These two tasks have different structural requirements for the lung – gas exchange requires thin walls, but the lung tissue must also be easily distortable to allow the lungs to increase and decrease volume to enable breathing. Optimizing the two opposing structural requirements means that over one's life, this likely contributes to an aging lung's weakening or deteriorating of the alveolar membrane (or for people with chronic respiratory conditions, like emphysema). It would be useful to know if exposure to ions exacerbates this.

Questions Regarding the Potential Impacts from Ionization Exposure

Given this framing of our cardiovascular autonomic nervous system (and cardiorespiratory system more broadly), we can start to develop formal questions informed by the different characterizations, such as the following:

1. Does repeated, long-term exposure to the ionization levels needed to be effective in removing pathogens and pollutants in real-world settings disrupt the cardiovascular autonomic nervous system's mechanisms and therefore human health? Here we have a question focused on understanding how the environmental conditions created by the ionization technology might impact the

mechanisms of the cardiovascular autonomic nervous system, and ultimately its ability to function as it was evolutionarily designed to do.

2. How does this exposure impact the development of the cardiovascular autonomic nervous system (and cardiorespiratory system more broadly) from conception to adulthood, and changes in how the system functions as a person ages past adulthood? Here we have a question focused on understanding how the development of the cardiovascular autonomic nervous system, as well as how it functions at different stages of development (influenced by the system's heritable history), are impacted by the environment created by ionization technology.

3. How do these levels compare to the levels our relevant evolutionary ancestors experienced in their environments (our ancestral environments)? Here we have a key question focused on understanding how the environment created by the ionization technology, particularly ionization concertation levels and time of exposure, differs from the relevant ancestral environment(s), which further informs how the first two questions need to be answered (which will be discussed further).

Starting with the first question, Table 4.1 lists the ranges of ionization levels encountered in a set of real-world studies looking at certain physiological impacts from exposure to the ions themselves. While one study after a few hours of exposure found a positive impact on the cardiovascular system's recovery after exercising [48], two studies found heart rate variability (HRV) indices were negatively altered after five days of exposure in primary classrooms [40,49], and a fourth study found indications of systemic oxidative stress after five days of exposure in student dormitories [41]. Oxidative stress can be associated with or

Table 4.1 The ranges of ionization levels encountered during a set of real-world studies looking at certain physiological impacts from exposure to the ions themselves.

Study Focus	Ion Concentration Levels	Impacts on Cardiovascular autonomic nervous system at levels needed to reduce indoor pollution levels
Classrooms, kids 5 days of exposure [41]	$12{,}997 +/- 3{,}814 cm^{-3}$ versus $12 +/- 10 cm^{-3}$	Heart rate variability (HRV) indices were negatively altered
Classrooms, kids 5 days of exposure [51]	$12{,}265 cm^{-3}$ versus $11 cm^{-3}$	HRV indices were negatively altered
Dormitories, university students, 5 days of exposure [52]	$60{,}591 +/- 12, 184 cm^{-3}$ versus $53 +/- 16 cm^{-3}$	Increases in biomarkers indicating systemic oxidative stress
Lab setting, adult men few hours of exposure [50]	$8{,}000 - 10{,}000 cm^{-3}$ versus $200 - 400 cm^{-3}$	Improved post-exercise recovery of cardiovascular & endocrine systems

cause cardiovascular autonomic dysfunction [50].These studies do suggest that multiple days of consistent exposure to Negative Air Ion (NAI) concentration levels high enough to have some effect at eliminating pathogens and pollutants can disrupt the mechanisms by which the cardiovascular autonomic nervous system operates. Additional research is needed to better understand the impacts after months or years of exposure versus only days of exposure, as well as greater understanding of the impacts on the systems' various mechanisms.

Regarding the second question involving the development of the cardiovascular autonomic nervous system (and cardiorespiratory system more broadly) over the course of an individual's lifetime, the author isn't aware of any studies specifically comparing the impacts of ion exposure on the systems of different age groups. But given the prior discussions, studying the impacts of ion exposure on the development of the lung's alveolar surface area through adulthood would be important to know, as would understanding age's influence on the apparent impacts of ion exposure on heart rate variability indices, oxidative stress, and cardiovascular system recovery after exercise. And if it was determined that ion exposure negatively impacts the efficiency of gas exchange across the lung's alveolar membrane, then further study would be necessary to understand how age influences this, including a focus on the lung's weakening or deteriorating of the alveolar membrane with age. There are obviously many other questions that could be asked.

Regarding the third question involving the nature of our ancestral environments' NAI concentration levels, we'll start with a comparison of the ion concentration levels found in contemporary natural environments with the ionization levels needed for this technology to be effective in real-world conditions. Table 4.2 provides a summary of the ranges of ion concentration levels found in various contemporary natural environments. The lowest concentration levels tend to be found at lake shores and in open space lacking vegetation. Concentration levels tend to be somewhat higher in forests. However, significantly higher concentration levels are found in association with waterfalls, rivers, mountain springs, and thunderstorms [51].

Table 4.2 The ranges of ionization levels encountered in selected contemporary natural environments.

Contemporary Natural Environment	*Ion Concentration Levels*
Lake Shores	$487cm^{-3}$ to $1,394cm^{-3}$ ($892cm^{-3}$ daily average) [53]
Forests	$1,001cm^{-3}$ to $5,515cm^{-3}$ ($2,871cm^{-3}$ daily average) [53]
Open space (no vegetation)	Similar to lake shores [53]
Waterfalls, rivers, mountain springs, thunderstorms	Tens of thousands per cubic cm [53]

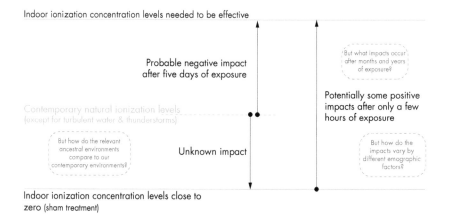

Figure 4.2 Summary of the potential impacts of ionization exposure on the cardiovascular autonomic nervous system covered in this chapter, comparing three different general NAI concentration levels (approximately zero, those of most contemporary natural environments, and the higher levels needed indoors to be effective for air purification). Image by Author.

Figure 4.2 summarizes what can be gathered from comparing these ranges of naturally occurring ion concentration levels in Table 4.2 to the indoor technology application ranges discussed here and summarized in Table 4.1. The NAI concentration levels encountered using this technology, as established in these studies, are higher than what humans typically experience within most contemporary natural environments. Referring to Tables 4.1 and 4.2, the lowest ion concentration levels from these referenced ionization studies range approximately from 6 to 26 times those found in lake shores and open spaces and 2 to 13 times those found in forests. It's significantly greater for the highest ion concentration levels from the referenced ionization studies. With waterfalls, rivers, mountain springs, and thunderstorms we start to see similar ion concentration levels or potentially even greater to those used in these studies.

Turbulent water[i], thunderstorms, and other natural events temporarily producing higher concentration levels of NAI are limited in how often they occur. Our exposure to these naturally occurring higher concentration levels is therefore limited. The open spaces lacking vegetation and forests are likely better reflective of the majority of outdoor time spent by most humans now, and by our ancestors in the past.

It should also be noted that the NAI concentration levels found in the natural world are significantly more variable than what are found indoors when an ionization system is operating. A lot of factors impact natural NAI concentration levels (and NAI "lifespans"), including landscapes, meteorological conditions, and the presence of various pollutants or compounds [51]. All of this could be one component of an explanation for why the study involving hours of exposure to ions at higher concentration levels had a positive impact while the studies involving days of exposure had negative impacts (though multiple other variables must be accounted for as well). This also suggests more studies are needed to compare the impacts from exposure to consistent NAI concentration levels high enough for pollutant removal to the impacts from conditions that better mimic the natural environment. If our cardiovascular autonomic nervous system evolved to most effectively function when exposed to the ranges of ion concentration levels found in contemporary natural environments (assuming this is an effective proxy for our ancestral environments) it is possible negative impacts could also happen at levels lower (or less variable) than what occurs in nature. **Only comparing the technology's ion concentration levels to the low levels when the equipment is off may not be an effective test in and of itself**.

However, we must also assess whether our contemporary natural environments are an effective proxy for our ancestral environments relative to NAI concentration levels. It's a complex question involving more than just the environments of modern humans and our immediate ancestors. We must travel back far enough along the evolutionary tree to fully understand the evolutionary development of the cardiovascular autonomic nervous system (and cardiorespiratory system more broadly) in relation to the relevant associated environmental conditions, including natural NAI concentration levels (as best as we can determine). So, for now it remains unknown how well contemporary natural environments act as proxies for ancestral environments relative to NAI exposure. Using it as a proxy should be done with caution. Additionally, it is sometimes claimed that exposure to ions shouldn't be harmful because they are naturally occurring [52]. But as has been discussed, at least regarding the functioning of the cardiovascular autonomic nervous system, the question of harm is likely dependent on the NAI concentration level, duration of exposure, and frequency of exposure. The nature of the exposure, how the impacts might be mediated by other aspects of the environment, age, pre-existing health conditions, etc., all shaped by our evolutionary history, must be considered before concluding NAI exposure isn't harmful.

Concluding Thoughts

Ionization technology, a type of emergent additive air cleaning technology, is sometimes proposed as a solution for cleaning our indoor air of pathogens and pollutants (sourced from the indoors or outdoors). Setting aside the potential negative impacts of the resulting air chemistry within the built environment, there is evidence that exposure to NAIs themselves may negatively impact human health, specifically the functioning of the cardiovascular autonomic nervous system (and cardiorespiratory system more broadly). But that impact is complex, in part due to our evolutionary history likely exposed to concentration levels of NAI lower and more variable than those needed to have some impact cleaning the air in our buildings. More research and literature reviews are needed, as laid out here, but it is a distinct possibility that the use of ionization technology creates an evolutionary mismatch for our cardiovascular autonomic nervous system and cardiorespiratory system more broadly. It's also possible that other physiological systems of traits might be negatively or positively impacted by exposure to NAI [53]. Some of this ionization technology also makes use of positive ions, whose impacts should also be further explored through additional research, and the concept of evolutionary mismatches provides a useful framework to structure the additional research needed. Referencing the previous chapter (Haines), the growing recognition of the importance of the built environment's microbiome to human health means we should also assess the impacts of ion exposure on the microbiome. And note that across the development spectrum, from conception through old age, we tend to pay less attention to the needs of people at either end. It's arguable that both ends have less of a voice, both in general and with respect to the design process in the AEC Industry. As a result, it's possible that these age ranges tend to experience a greater number of evolutionary mismatches and or are subjected to their more severe impacts. Adopting an evolutionary framework as laid out here can increase the *respiratory equity* (Vance) within our built environments.

Recognizing that human traits (physiological, psychological, and social/cultural) may not be optimally suited for contemporary environments – how we design them, how we operate them, what technology we integrate within them, etc. – because of our evolutionary history provides us with an opportunity to understand why that is the case and better optimize our environments based on that understanding. Tinbergen's four trait characterizations allow us to develop questions focused on understanding the complex, nuanced nature of how traits are mismatched with contemporary, contextual environments, as well as the potential short and long-term negative impacts of such mismatches, further improving productivity, health, and ultimately our evolutionary fitness for individuals, the groups they're a part of, and the building/occupant organism as a whole.

Note

i   The high concentration levels of ions at these bodies of water are specifically associated with turbulent water and are a result of the Lenard effect (the separation of electric charges accompanying the aerodynamic breakup of water drops) [51]. But it's important to note that such turbulent waters are limited in where they occur (and therefore how often we are exposed to them) in both contemporary and ancestral environments.

References

[1] Harmon M. Constructing Our Niches: The Application of Evolutionary Theory to the Architecture, Engineering, and Construction (AEC) Industry. *This View of Life*; September 12, 2018. Accessed September 17, 2022. https://thisviewoflife.com/

constructing-our-niches-the-application-of-evolutionary-theory-to-the-architecture-engineering-and-construction-aec-industry/

[2] Henrich J. *The Secret of Our Success: How Culture Is Driving Human Evolution, Domesticating Our Species, and Making Us Smarter.* Princeton University Press; 2016.

[3] Wilson DS. *Does Altruism Exist? Culture, Genes, and the Welfare of Others.* Yale University Press; 2015.

[4] Li NP, van Vugt M, Colarelli SM. The Evolutionary Mismatch Hypothesis: Implications for Psychological Science. *Current Directions in Psychological Science.* 2017;27(1):38–44. doi:10.1177/0963721417731378

[5] Giosan C, Muresan V, Wyka K, Mogoase C, Cobeanu O, Szentagotai A. The Evolutionary Fitness Scale: A Measure of the Independent Criterion of Fitness. *The Journal of the Evolutionary Studies Consortium.* 2018;8(1):1–28. Accessed November 7, 2022. https://evostudies.org/volume-8/

[6] Sober E. *Did Darwin Write the Origin Backwards? Philosophical Essays on Darwin's Theory.* Prometheus Books; 2011.

[7] Richerson P, Baldini R, Bell AV, et al. Cultural Group Selection Plays an Essential Role in Explaining Human Cooperation: A Sketch of the Evidence. *Behavioral and Brain Sciences.* 2016;39:e30. doi:10.1017/S0140525X1400106X

[8] Uchiyama R, Spicer R, Muthukrishna M. Cultural Evolution of Genetic Heritability. *Behavioral and Brain Sciences.* 2022;45:e152. doi:10.1017/S0140525X21000893

[9] Micheletti AJC, Brandl E, Mace R. What Is Cultural Evolution Anyway? *Behavioral Ecology.* 2022;33(4):667–669. doi:10.1093/beheco/arac011

[10] Wilson DS, Ostrom E, Cox ME. Generalizing the Core Design Principles for the Efficacy of Groups. *Journal of Economic Behavior and Organization.* 2013;90:S21–S32. doi:10.1016/j.jebo.2012.12.010

[11] Sober E. *The Nature of Selection: Evolutionary Theory in Philosophical Focus.* The University of Chicago Press; 1993. Accessed September 19, 2022. https://press.uchicago.edu/ucp/books/book/chicago/N/bo3621916.html

[12] Cedeño Laurent JG, Macnaughton P, Jones E, et al. Associations Between Acute Exposures to PM2.5 and Carbon Dioxide Indoors and Cognitive Function in Office Workers: A Multicountry Longitudinal Prospective Observational Study. *Environmental Research Letters.* 2021;16(9):094047. doi:10.1088/1748-9326/AC1BD8

[13] Allen JG, MacNaughton P, Satish U, Santanam S, Vallarino J, Spengler JD. Associations of Cognitive Function Scores with Carbon Dioxide, Ventilation, and Volatile Organic Compound Exposures in Office Workers: A Controlled Exposure Study of Green and Conventional Office Environments. *Environmental Health Perspectives.* 2016;124(6):805–812. doi:10.1289/ehp.1510037

[14] Dai D, Prussin AJII, Marr LC, Vikesland PJ, Edwards MA, Pruden A. Factors Shaping the Human Exposome in the Built Environment: Opportunities for Engineering Control. *Environmental Science & Technology.* 2017;51(14):7759–7774. doi:10.1021/acs.est.7b01097

[15] Nezis I, Biskos G, Eleftheriadis K, Kalantzi OI. Particulate Matter and Health Effects in Offices – A Review. *Building and Environment.* 2019;156:62–73. doi:10.1016/j.buildenv.2019.03.042

[16] McArthur JJ, Powell C. Health and Wellness in Commercial Buildings: Systematic Review of Sustainable Building Rating Systems and Alignment with

Contemporary Research. *Building and Environment*. 2020;171:106635. doi:10.1016/j. buildenv.2019.106635

[17] Tham KW, Willem HC. Room Air Temperature Affects Occupants' Physiology, Perceptions and Mental Alertness. *Building and Environment*. 2010;45(1):40–44. doi:10.1016/j.buildenv.2009.04.002

[18] Liu C, Zhang Y, Sun L, Gao W, Jing X, Ye W. Influence of Indoor Air Temperature and Relative Humidity on Learning Performance of Undergraduates. *Case Studies in Thermal Engineering*. 2021;28:101458. doi:10.1016/j.csite.2021.101458

[19] Lan L, Tang J, Wargocki P, Wyon DP, Lian Z. Cognitive Performance Was Reduced by Higher Air Temperature Even When Thermal Comfort Was Maintained Over the 24–28°C Range. *Indoor Air*. 2022;32(1):e12916. doi:10.1111/ina.12916

[20] Wargocki P, Wyon DP. Ten Questions Concerning Thermal and Indoor Air Quality Effects on the Performance of Office Work and Schoolwork. *Building and Environment*. 2017;112:359–366. doi:10.1016/j.buildenv.2016.11.020

[21] Yang L, Yan H, Lam JC. Thermal Comfort and Building Energy Consumption Implications – A Review. *Applied Energy*. 2014;115:164–173. doi:10.1016/j. apenergy.2013.10.062

[22] Kim J, Bauman F, Raftery P, et al. Occupant Comfort and Behavior: High-Resolution Data from a 6-Month Field Study of Personal Comfort Systems with 37 Real Office Workers. *Building and Environment*. 2019;148:348–360. doi:10.1016/ j.buildenv.2018.11.012

[23] Lloyd E, Sloan Wilson D, Sober E. *Evolutionary Mismatch and What to Do About It: A Basic Tutorial*; 2011. Accessed April 5, 2021. http://mypage.iu.edu/~ealloyd/ http://evolution.binghamton.edu/dswilson/3http://philosophy.wisc.edu/sober/

[24] Wilson DS, Basile AJ, Smith JB. Evolutionary Mismatch and What to Do About It – Evolutionary Mismatch Series. *This View of Life Magazine*; 2019. Accessed April 5, 2021. https://thisviewoflife.com/evolutionary-mismatch-and-what-to-do-about-it-2/

[25] Wilson DS. *This View of Life: Completing the Darwinian Revolution*. First ed. Pantheon; 2019. Accessed April 18, 2021. www.harvard.com/book/this_view_ of_life_completing_the_darwinian_revolution/

[26] Vetter C, Pattison PM, Houser K, et al. A Review of Human Physiological Responses to Light: Implications for the Development of Integrative Lighting Solutions. *LEUKOS*. 2022;18(3):387–414. doi:10.1080/15502724.2021.1872383

[27] Figueiro MG, Nagare R, Price LLA. Non-Visual Effects of Light: How to Use Light to Promote Circadian Entrainment and Elicit Alertness. *Lighting Research & Technology*. 2017;50(1):38–62. doi:10.1177/1477153517721598

[28] Brown TM, Brainard GC, Cajochen C, et al. Recommendations for Daytime, Evening, and Nighttime Indoor Light Exposure to Best Support Physiology, Sleep, and Wakefulness in Healthy Adults. *PLOS Biology*. 2022;20(3):e3001571. doi:10.1371/ journal.pbio.3001571

[29] Klepeis NE, Nelson WC, Ott WR, et al. The National Human Activity Pattern Survey (NHAPS): A Resource for Assessing Exposure to Environmental Pollutants. *Journal of Exposure Science and Environmental Epidemiology*. 2001;11(3):231–252. doi:10.1038/sj.jea.7500165

[30] Stephens B, Gall ET, Heidarinejad M, Farmer KK. Interpreting Air Cleaner Performance Data. *ASHRAE Journal*. 2022;64(4):20–30. Accessed December 8, 2022. https://technologyportal.ashrae.org/journal/articledetail/2392

[31] EMG (Environmental and Modelling Group). *EMG: Potential Application of Air Cleaning Devices and Personal Decontamination to Manage Transmission of COVID-19*; November 4, 2020. Accessed March 8, 2021. www.gov.uk/government/publications/emg-potential-application-of-air-cleaning-devices-and-personal-decontamination-to-manage-transmission-of-covid-19-4-november-2020

[32] Ratliff KM, Oudejans L, Archer J, et al. Large-Scale Evaluation of Microorganism Inactivation by Bipolar Ionization and Photocatalytic Devices. *Building and Environment.* 2023;227:109804. doi:10.1016/J.BUILDENV.2022.109804

[33] Zeng Y, Manwatkar P, Laguerre A, et al. Evaluating a Commercially Available In-Duct Bipolar Ionization Device for Pollutant Removal and Potential Byproduct Formation. *Building and Environment.* 2021;195:107750. doi:10.1016/j.buildenv.2021.107750

[34] Zeng Y, Heidarinejad M, Stephens B. Evaluation of an In-Duct Bipolar Ionization Device on Particulate Matter and Gas-Phase Constituents in a Large Test Chamber. *Building and Environment.* 2022;213:108858. doi:10.1016/J. BUILDENV.2022.108858

[35] Licht S, Hehir A, Trent S, et al. *Use of Bipolar Ionization for Disinfection Within Airplanes*; 2021. Accessed May 11, 2021. www.boeing.com/confident-travel/downloads/Boeing-Use-of-Bipolar-Ionization-for-Disinfection-within-Airplanes.pdf

[36] Collins DB, Farmer DK. Unintended Consequences of Air Cleaning Chemistry. *Environmental Science & Technology.* 2021;55(18):12172–12179. doi:10.1021/acs. est.1c02582

[37] Zannoni N, Lakey PSJ, Won Y, et al. The Human Oxidation Field. *Science (1979).* 2022;377(6610):1071–1077. doi:10.1126/science.abn0340

[38] Ault AP, Grassian VH, Carslaw N, et al. Indoor Surface Chemistry: Developing a Molecular Picture of Reactions on Indoor Interfaces. *Chem.* 2020;6(12):3203. doi:10.1016/J.CHEMPR.2020.08.023

[39] Joo T, Rivera-Rios JC, Alvarado-Velez D, Westgate S, Ng NL. Formation of Oxidized Gases and Secondary Organic Aerosol from a Commercial Oxidant-Generating Electronic Air Cleaner. *Environmental Science & Technology Letters.* 2021;8(8):691–698. doi:10.1021/acs.estlett.1c00416

[40] Dong W, Liu S, Chu M, et al. Different Cardiorespiratory Effects of Indoor Air Pollution Intervention with Ionization Air Purifier: Findings from a Randomized, Double-Blind Crossover Study Among School Children in Beijing. *Environmental Pollution.* 2019;254:113054. doi:10.1016/j.envpol.2019.113054

[41] Liu W, Huang J, Lin Y, et al. Negative Ions Offset Cardiorespiratory Benefits of PM2.5 Reduction from Residential Use of Negative Ion Air Purifiers. *Indoor Air.* 2021;31(1):220–228. doi:10.1111/ina.12728

[42] Tinbergen N. On Aims and Methods of Ethology. *Zeitschrift für Tierpsychologie.* 1963;20(4):410–433. doi:10.1111/j.1439-0310.1963.tb01161.x

[43] Wilson DS. Small Groups as Fundamental Units of Organization. In: Wilson DS, Hayes SC, eds. *Evolution and Contextual Behavioral Science.* Context Press; 2018:245–259.

[44] Hägglund H, Uusitalo A, Peltonen JE, et al. Cardiovascular Autonomic Nervous System Function and Aerobic Capacity in Type 1 Diabetes. *Frontiers in Physiology.* 2012;3. doi:10.3389/FPHYS.2012.00356

[45] Zeman KL, Bennett WD. Growth of the Small Airways and Alveoli from Childhood to the Adult Lung Measured by Aerosol-Derived Airway Morphometry. *Journal of Applied Physiology.* 2006;100(3):965–971. doi:10.1152/japplphysiol.00409.2005

[46] Sharma G, Goodwin J. Effect of Aging on Respiratory System Physiology and immunology. *Clinical Interventions in Aging.* 2006;1(3):253–260. doi:10.2147/ciia.2006.1.3.253

[47] West JB, Watson RR, Fu Z. The Human Lung: Did Evolution Get It Wrong? *European Respiratory Journal.* 2007;29(1):11–17. doi:10.1183/09031936.00133306

[48] Ryushi T, Kita I, Sakurai T, et al. The Effect of Exposure to Negative Air Ions on the Recovery of Physiological Responses After Moderate Endurance Exercise. *International Journal of Biometeorology.* 1998;41(3):132–136. doi:10.1007/s004840050066

[49] Liu S, Huang Q, Wu Y, et al. Metabolic Linkages Between Indoor Negative Air Ions, Particulate Matter and Cardiorespiratory Function: A Randomized, Double-Blind Crossover Study Among Children. *Environment International.* 2020;138:105663. doi:10.1016/j.envint.2020.105663

[50] Ziegler D, Buchholz S, Sohr C, Nourooz-Zadeh J, Roden M. Oxidative Stress Predicts Progression of Peripheral and Cardiac Autonomic Nerve Dysfunction Over 6 Years in Diabetic Patients. *Acta Diabetologica.* 2015;52(1):65–72. doi:10.1007/s00592-014-0601-3

[51] Wang H, Wang B, Niu X, et al. Study on the Change of Negative Air Ion Concentration and Its Influencing Factors at Different Spatio-Temporal Scales. *Global Ecology and Conservation.* 2020;23:e01008. doi:10.1016/J.GECCO.2020.E01008

[52] Walker D, Browning W. *The Nature of Air: Economic & Bio-Inspired Perspectives on Indoor Air Quality Management*; 2019. Accessed December 13, 2022. www.terrapinbrightgreen.com/report/the-nature-of-air/

[53] Jiang SY, Ma A, Ramachandran S. Negative Air Ions and Their Effects on Human Health and Air Quality Improvement. *International Journal of Molecular Sciences.* 2018;19(10). doi:10.3390/ijms19102966

5 Designing with Metrics for Indoor Air Quality, Comfort, and Health

Z Smith

Great architecture, whether humble or grand, resists quantification. Yet architecture is built on numbers: The beam is either large enough to support the floor load or it is not, and that's something we can quantify. It doesn't lessen the beauty of a building to quantify whether the beam is adequately sized, or to quantify other key aspects of the environment it provides for those who use it. James Marston Fitch opened *American Building: The Environmental Forces that Shape It* with the assertion that "the ultimate task of architecture is to act in favor of human beings – to interpose itself between people and the natural environment . . . to remove the gross environmental load from their shoulders" [1]. He lamented that while all architects aspire to the creation of beautiful buildings, many focus on aesthetics first over the rest of human experience. He argued that meeting the very human needs for light, air, and comfort is as central to good architecture as beautiful imagery. While what qualifies as "beautiful imagery" will likely remain somewhat difficult to put to a quantitative test, we can become fluent in the metrics of indoor air quality, comfort, and health and use these metrics to inform design.

Quantifying Comfort and the "Standard Environment"

Let's begin with the standard used by mechanical engineers to quantify thermal comfort: ASHRAE Standard 55, "Thermal Environmental Conditions for Human Occupancy". This standard was developed using voting to link the subjective notion of human comfort to the objective metrics of the temperature, relative humidity, and velocity of air, as well as the surface temperatures of the enclosing walls, floors, and ceiling. This correlation was established through research in which groups of test subjects dressed in a range of clothing and sitting or performing light work assessed their comfort level on a 5-point scale as the indoor conditions were varied. For each set of circumstances, occupants were also asked to describe themselves as being "satisfied" or "dissatisfied" with their environment. What's interesting is that there was no single set of conditions for which 100% of the test subjects said they were "satisfied" – the conditions that

DOI: 10.1201/9781003398714-6

one individual found comfortable were too cool or too warm for someone else wearing the same level of clothing and at the same level of activity. Instead, the standard is defined such that a space achieving conditions where 80% of the occupants would report being "satisfied" is deemed as acceptable. That means a space delivering conditions where 20% of the occupants are dissatisfied is still rated as meeting the standard for comfort!

Researchers have observed this "one size never fits all" truth from the beginning. Beginning with Ole Fanger's work in the 1960s to quantify human comfort in buildings, it has been observed that the conditions providing the greatest comfort to one individual will be different than those of another [2]. While the earliest studies thought that differences among individuals could be explained through differences in size, weight, clothing level, and the like, more recent research has demonstrated that even with all those conditions being equal, people of different genders will report differently on comfort [3,4]. For example, women often report feeling uncomfortably cold at temperatures where men report being satisfied with the temperature, and these differences by gender can have big impacts for equity in the workplace. A recent series of very striking experiments found that the temperature at which women score best in math and verbal tests is significantly warmer than the temperature at which men score best [5]. If the thermostat is set by the manager, and more managers are men, then if they set the temperature at which they themselves are comfortable, they will disadvantage the performance of women. Nevertheless, the technical experts on the ASHRAE Research Administration Committee (RAC) assert that the differences between temperature and comfort are solely traceable to differences in clothing level preferred by men and women.

Historically, people have adapted their clothing and their diet to the seasons, consistent with indoor temperatures being lower in winter and higher in summer. This observation was quantified through research into surveys of user comfort in various indoor temperatures in climates across the globe by Richard de Dear and Gail Brager, who determined that interior comfort would vary with outdoor temperature conditions [6]. This was incorporated into the ASHRAE 55 thermal comfort standard beginning in 2004 as the "Adaptive Comfort Standard" (ACS) as an alternative to the traditional range of temperatures. Amusingly, ASHRAE held firm that this approach was only applicable for "occupant-controlled naturally conditioned spaces . . . where the thermal conditions of the space are regulated primarily by the occupants through opening and closing of windows" [7]. No explanation was given for why this same observation would not apply for fully mechanically conditioned buildings.

Fascinatingly, there is mounting evidence that even for one individual, it may be better for their health to vary the temperature over the course of the day than to maintain an "ideal" fixed temperature [8], even if these temperature excursions are well outside the normal comfort zone. In fact, the term "thermal alliesthesia" (alliesthesia meaning literally "change-sensing") is used to describe the pleasure experienced by changes in temperature [9]. This will come as no surprise to the readers of Lisa Heschong's classic *Thermal Delight in Architecture*, where the author made the case that it is the variability in thermal environment – the radiant glow of a campfire on a cold night, the beam of direct sun warming one through the window on a cold winter's day, the plunge into an icy lake after a sauna – that can create our most memorable and pleasurable thermal experiences [10]. This variation among individuals, and the need for variation over time even for one individual, is for more than just temperature. The level of lighting (illuminance) recommended by standards organizations such as the Illuminating Engineering Society of North America (IESNA) vary depending both on the visual task and the age of the person [11]. We cannot simply oversupply light to meet the requirements of those needing the highest light levels, because the same lighting that may be the minimum needed for the elderly may be perceived as uncomfortably bright by children at the same library. Moreover, studies show that spaces illuminated through natural daylight, which varies over the course of the day, often yield higher occupant satisfaction, retail sales, worker productivity, and student learning rates than spaces illuminated to the same levels with electric lighting [12]. In part, this can be attributed to the benefit of the variability itself.

So, we come back to this: People are different, and their differences in preferred indoor conditions have measurable consequences for cognitive performance, productivity, and health. And even for any one individual, the "optimal" conditions may vary according to the activity, state of mind, and time of day. Instead, it may be desirable not to provide even that one individual with this "optimal" set of conditions, but to let them vary. Yet there is a tendency for some engineers to deliver the "standard environment": 72°F, 50% relative humidity, 30 foot-candles horizontal illuminance. While this pursuit of the *least-offensive* indoor environment may minimize complaints, it will never eliminate them, nor will it unlock the best of what we can be. Drawing from the attributes of great architecture cited by Vitruvius as "Strength, Utility, and Delight", we need to design our indoor environments to aim for both *utility* – meeting the minimal needs of the individual – and *delight*, providing the variability that stimulates and energizes us.

Strategies for Passive, Active, and *Smart* Design

Given that outdoor temperatures can vary greatly over the course of the day and the season, the task that Fitch has set for designers of buildings is to provide indoor conditions that temper those swings, while allowing for the variety and delight that Heschong described. It is common to characterize these approaches as either "passive" (those using building materials, form, and orientation to do much of the work of providing thermal comfort, air, and light) or "active" (those using energy-consuming systems to deliver or extract heat, move air, and provide artificial light). We could propose a middle way, "smart", which starts with passive approaches and deploys active approaches only as needed. Not every climate nor every program will allow for a completely passive approach. For example, if temperatures over the course of the day stay consistently well outside the range of human comfort, even after skillful deployment of passive strategies, then active approaches may be required. However, by starting with passive approaches first and then overlaying active strategies, we can provide for more human comfort with the least energy [13]. That's smart.

Operable Windows

The unsung hero of smart design is the humble operable window. The origin of the word itself, drawn from the words for 'wind' and 'eye', captures its dual role in ventilation and light, with the implicit notion of operability. Before the development of active mechanical heating and cooling systems, the window was the key element of how occupants controlled the indoor environment – opening to let in fresh air and cooling breezes as desired, shut and covered with insulating cloth to block the cold or heat when not. How people use operable windows depends in part on how a building is heated. In the late 19th century, when open fireplaces fed with wood or coal were the common sources of heat, awareness grew of the health dangers of stale, particulate-laden air; abolitionist and reformer Harriet Beecher Stowe became an advocate for the use of through sealed-combustion Franklin stoves while leaving the windows open [14]. When the role of airborne transmission of diseases such as influenza became understood in the early 20th century, the window's role in the provision of fresh air became paramount, as discussed in Chapter 1, and one saw the proliferation of heating systems that made up for the larger quantities of cool outdoor air flowing through indoor spaces via steam radiators [15–17] or radiant floors or ceilings [18]. Rather than sealing the building and heating the air, these strategies allowed for abundant cooler outdoor air while using radiative sources to provide thermal balance for occupants. This strategy is an ancient one, dating back to the hypocausts of Roman Republic and the Korean *ondol* of 1000 BCE [19]. Operable windows enjoyed renewed interest at the height of the COVID-19 pandemic. Whatever the building's mechanical system, opening the windows offered a way to greatly increase the fraction of outdoor air and lower the transmission risk of airborne diseases [20].

Yet today, sealed buildings conditioned by air-based systems rule commercial construction. How did we get here? Air conditioning. More specifically, dehumidification. It's important to recall that Willis Carrier's 1902 invention of the first modern air conditioning system was actually designed as a means of controlling indoor humidity levels to maintain industrial machinery [21]. Humans maintain thermal balance with their environment through a combination of evaporation of water through the skin as well as radiative transfer to the surrounding environment and convective coupling to surrounding air [19]. The more humid the surrounding air, the less quickly our bodies can cool via evaporation. Raising the rate at which air flows over our bodies (through fans or natural ventilation strategies) can help, but there are practical limitations to this approach. Moreover, elevated indoor humidity levels can promote the growth of mold, leading to negative health impacts, and so the EPA recommends indoor relative humidity levels be kept below 60%, ideally below 50% [22]. Now, we don't have to seal our building all the time, only when the moisture content of outdoor air is sufficiently high to result in indoor relative humidity levels above 60%. We can still have operable windows in buildings and maintain humidity control in climates with high humidity so long as we close them during these periods of high humidity. Even in famously humid climates of the American Southeast, favorable conditions apply to thousands of operating hours per year.

Once one has provided ducts to deliver humidity-controlled air to a building during *some* periods, the temptation is to simply let it manage conditions *all* the time. Once we've installed all the ductwork for air conditioning and ventilation, why have a separate system for heating? Hence the emerging dominance of the forced ducted air heating & cooling system since the 1960s. Using air as the means for moving heat around in buildings, where a fan at a central unit pressurizes the air to move it through ducts and draws air back through a return system via this pressure differential means that the systems in fact rely upon the building being sealed. Nothing strikes fear into a mechanical engineer of an air-based HVAC system like an open window; it will "unbalance" the system . . . or the engineer.

Interestingly, engineers aren't opposed to using large quantities of outdoor air to maintain temperature when conditions are right; they just give this approach a special name: economizer mode [23]. In fact, in all but the hottest and most humid climate zones, modern commercial building energy codes, such as the International Energy Conservation Code (IECC), require support for economizer mode. A building with a forced-air HVAC system can still have operable windows; it just needs to communicate with the occupants that conditions are right and let the occupants open the windows [24]. Operable windows then can be the ultimate economizer mode. Similarly, the WELL building standard encourages operable windows to be provided with indicators encouraging them to be opened depending on outdoor air temperature, relative humidity, and particulate matter levels [25].

But do operable windows actually save energy? *It depends* [26–28]. If outdoor air temperature and humidity levels are favorable, then operable windows can allow fresh air to be introduced and temperature controlled without requiring fans pushing air around inside ducts. Fan energy can comprise 10–20% of commercial building energy consumption, depending on climate and duct design. However, these moderate outdoor conditions are exactly those requiring the least amount of heating and cooling, so savings may be modest. Worse, windows may be left open during unfavorable conditions while the HVAC system continues operation, so that system can pump conditioned air directly outdoors. Intelligent systems and engaged users are key to success, but if part of the mission of architecture is to provide not mere functionality but *delight*, then throwing open the windows on a beautiful spring day is an experience we shouldn't deny occupants.

The Porch

One architectural device that allows people to "vote with their feet" and move to where they are most comfortable is the porch. This transitional space, which offers protection from above while being open to the outdoors on most sides, is neither truly indoors nor truly outdoors. The porch is an in-between space, not just thermally but socially, and visitors might be welcomed to a porch who are not invited inside. It offers both a view to the outdoors and some degree of concealment – both "refuge" and "prospect," in the terms coined by Jay Appleton [29]. In vernacular architecture without mechanical air conditioning, porches may offer superior thermal comfort when spaces indoors may have built up heat from the day while the porch offers access to higher airflow or even a cool evening breeze. But even when thermal conditions on a porch, atrium, or other semi-outdoor environment do not meet the standard ASHRAE thermal comfort ranges, research shows that occupants are comfortable over a much wider range of temperatures [30], perhaps willing to trade the standard environment for the richness of the visual and sensory experience these spaces offer. It's as if the ASHRAE 55 Adaptive Comfort Standard takes hold the instant one steps out from the fully conditioned interior.

Filtration

Particulates in outdoor air can come from combustion exhaust from power plants or motor vehicles, or from wildfires. In such conditions, filtration of outdoor air becomes essential. Buildings located in regions used to high-quality outdoor air have sometimes been unprepared for the challenges forced by the increasing frequency and severity of wildfires in the American west. During wildfires in California, for example, some office buildings had HVAC systems with no accommodations for air intake filters, leading operators to reduce or shut off air intake during severe periods [31,32]. In most recirculating forced-air-based HVAC systems, a filter is incorporated in the Air Handling Unit (AHU), typically just before the air hits the heating or cooling coils. These filters were introduced largely for the protection of the equipment, as the buildup of lint, dust, and other particles on coils can reduce heat transfer efficiency. Over time, it has become recognized that these filters could also potentially improve indoor air quality by trapping particulates. Environmental and health-oriented building rating systems such as LEED and WELL reward buildings providing high levels of filtration, typically characterized by the MERV (Minimum Efficiency Reporting Value) rating of the filter [33]. MERV 8 is the mandatory minimum and MERV 13 filtration is rewarded.

This higher level of filtration has been shown to lower the spread of airborne diseases, such as influenza, with benefits accruing for higher MERV ratings that flatten out above MERV 13 [34]. While viruses are much smaller than the 1 micron blocked by a MERV 13 filter, most viruses 'hitchhike' on small, micron-sized water droplets or aerosols produced as people exhale, and these aerosols can stay suspended in air for hours on end. As air is drawn back into the air handling unit, the filters can catch and trap these aerosols and each recirculating air change gives the filter another chance at trapping the aerosol. During the early days of the COVID-19 pandemic it was assumed that, like other coronaviruses, the dominant mode of transmission would be through direct contact of surfaces (hence all the early advice for handwashing) or large droplets emitted by coughing that would fall to the ground within a few feet (hence the six-foot social distancing). However, subsequent analysis of "superspreader" events where the only plausible mechanism was aerosol transmission led to recommendations to open windows, increase ventilation rates, and increase air filtration, either through high-MERV filters within recirculating air HVAC systems or in-room portable air cleaners [35,36]. This meant that designers arguing for operable windows and higher ventilation and filtration rates, motivated by the goal of increased occupant satisfaction and productivity, now had a new ally: the fear of the spread of a deadly disease. Recently, both the Centers for Disease Control and ASHRAE announced a way to quantify the impact of using ventilation, filtration, and UV sterilization in combination, a term they have dubbed the "Equivalent Outdoor Air Change Rate", and recommend an equivalent air change rate of 5 ACH (Air Changes per Hour) for designers seeking to minimize the spread of airborne diseases. This 5 ACH can be delivered by fresh air alone, or at much lower flows of fresh air used in combination with filtration and/or UV [37,38]. Additionally, the LEED green building rating system now offers a pilot credit for meeting this target [39]. Two health-based building certification systems, WELL and Fitwel, provide a roadmap to other key strategies, informed by peer-reviewed research. Providing good indoor air quality and thermal comfort are important components of an indoor environment supportive of occupant health and productivity. But they are not the whole story.

Fresh Air, Carbon Dioxide, and Bioeffluents

Humans breathe – it means we're alive. We inhale to take in oxygen which our bodies metabolize with our food to produce heat and energy, and we exhale to expel the carbon dioxide (CO_2) that is a byproduct of that metabolizing. If our buildings were perfectly sealed boxes, our exhaled CO_2 would build up indoors to levels that would at first render us sluggish, then sleepy, then dead. This is true even if there is still plenty of oxygen available [40,41]. A great illustration of this is the experience of the Apollo 13 lunar mission where, after an explosion, the spacecraft still had plenty of oxygen but the filtration system that removed excess CO_2 from the lunar lander (in which the astronauts had taken refuge) was insufficient to support all three astronauts for an extended period of time. The lives of the crew were saved by improvising a system to use the CO_2 scrubber from a different spacecraft module using, of course, duct tape [42].

In a sealed room or building, we avoid building up CO_2 by deliberately drawing in outdoor air while expelling a comparable amount of stale air. The CO_2 level will build up to a level above outdoor CO_2 level by simply the ratio of the rate at which an occupant exhales CO_2 to the rate at which fresh air is being drawn in [43]. For example, if a seated occupant exhales 0.010 cubic feet per minute (CFM) of CO_2, and outdoor air is drawn it at 10 CFM, then indoor CO_2 levels will be (0.010/20 = 0.0005) or 500 ppm above the CO_2 levels outdoors. If outdoor air is drawn in at twice the rate (40 CFM), then indoor air will stabilize at half that increment (250ppm) above outdoor levels. Given outdoor CO_2 levels of around 400ppm, then indoor levels at these two levels of ventilation would be 400 + 500 = 900 ppm and 400 +250 = 650ppm, respectively.

As recently as the late 1990s, indoor CO_2 levels below levels of 2000 ppm were not generally thought to be a problem, and indoor levels of 1000 ppm were typical. Rather, CO_2 levels were taken as a convenient way to measure the adequacy of ventilation of occupied buildings [44], where the real concern was human-generated odors, or "bioeffluents", as they were discreetly referred to in the ASHRAE ventilation standard [45]. Since the source of both the bioeffluents and CO_2 levels was people and CO_2 was relatively easy to measure, demand-controlled ventilation (DCV) systems could ramp up and down the fresh air and exhaust based on CO_2 readings to reduce unnecessary over-ventilation and thereby save the energy associated with dehumidifying, cooling, or warming outdoor air. Over time, energy codes required the use of DCV, but only in heavily ventilated assembly spaces such as auditoria or large classrooms.

Beginning in 2012, researchers from the State University of New York and Lawrence Berkeley National Lab conducted a study that there was a measurable difference in cognitive performance between 1000ppm and 2500ppm [46]. Building on this work, a team from the Harvard T.H. Chan School of Public Health and their collaborators published results indicating that the performance of office workers on tests of cognitive function improved markedly when indoor CO_2 levels and VOCs were significantly lower than those found in typical US office buildings [20,47]. Their experiments explored the impact of VOCs and CO_2 levels separately by first varying VOC concentrations while keeping ventilation rates and CO_2 levels at conditions found in typical offices. They saw significant test score improvements over a wide range of knowledge worker tasks when VOC levels were lowered from 500–700 micrograms per cubic meter to less than 50. Allen et al. found that the selection of low- or zero-VOC indoor finishes can result in significantly lower VOC levels even at standard ventilation rates. Even greater improvements in cognitive function scores were observed in low-VOC environments when ventilation rates were doubled (from 20 to 40 CFM per occupant) and CO_2 levels were lowered from 950 ppm to 600 ppm. Subsequent work showed that the relationship between CO_2 concentrations and cognitive function performance was linear over this range [48]. The Laurent study also investigated the impact of higher levels of particulate matter of dimensions 2.5 microns or less (PM2.5). The study noted that indoor PM2.5 levels in found in offices vary widely by country, ranging from 1 micron per cubic meter in the US to 18 in China, and they found that, all other things being held equal, high PM2.5 levels are associated with lower cognitive performance. While these studies are compelling, they have been met with some resistance. In 2022, the ASHRAE Board of Directors approved a position document declaring that "existing evidence for direct impacts of CO_2 on health, well-being, learning outcomes, and work performance at commonly observed indoor concentrations is inconsistent, and therefore does not currently justify changes to ventilation and IAQ standards, regulations, or guidelines" [49].

These studies are interesting because they are trying to measure mental performance as impacted by indoor air quality alone. The financial impact of this is potentially huge. For many firms, salaries make up the vast majority of expenses, and so elements that improve worker productivity by just 2 percent can translate to at least 1 percent more output and thus profit. From their analysis of salary and employment data showing how much more the market is willing to pay better-performing employees within a given job title, MacNaughton et al. (2015) argued that the improvement in performance demonstrated for knowledge workers when CO_2, VOC, and PM2.5 levels are lower correlates with those paid $6,500/year higher salaries. They then compare this with the energy costs associated with higher levels of ventilation (in order to lower CO_2 levels) in a variety of climates across the United States and find these costs to range from $1/year to $40/year, depending on climate and whether energy recovery technologies are employed. On balance, they argue that a higher ventilation rate (around 40 cfm per occupant rather than around 20 cfm typically used to meet the ASHRAE 62 ventilation standard) pays for itself many times over in worker productivity. If we use Allen et al.'s productivity analysis research (2016) and assume just a 1% increase in productivity for various improvements to visual comfort, the net financial benefits would be significant. Even a 1% gain in productivity for a knowledge worker whose salary and benefits total $100,000/year constitutes a gain of $1,000/year, far in excess of the maximum cost of increased ventilation of $40/year per employee found by Allen et al.

Long before Allen et al., researchers looked at the impact of indoor air quality on health, as quantified through the simple metric of sick days. In 2000, a team of Harvard researchers analyzed records for a single employer with 40 facilities and found that those with ventilation rates in the range of 50 CFM/person had significantly lower complaints concerning indoor air quality and significantly lower rates of employees taking sick leave than facilities with ventilation rates of approximately 25 CFM/person, both above the levels more typically found in facilities following the ASHRAE 62 ventilation standard. They found that the cost of providing this higher ventilation rate was one-sixth of the cost of the higher rates of sick leave [50]. These data were then used by those studying the use of economizers to make the case of a double benefit of both energy savings and sick leave savings [23].

Similarly, most green-building rating systems reward buildings that provide adequate daylight and views to the outdoors, especially views to nature. Heschong (2002) summarizes some of the most compelling research on the impact of daylight student performance, worker productivity and on retail sales. These results include improvements to student test speeds in daylit classrooms [51], enhanced performance in office workers with views [52], and increased sales in daylit stores [53]. A 2011 study described an analysis of the anonymized health records of employees in a university administrative services building when compared with the daylight and quality of the views from their desks. It found that those seated at desks with the worst access to daylight took an average of 2 more sick days per year, and those with the worst views (as quantified by a survey of pairwise rankings of photographs of the views from those desks) took an average of 1.4 more sick days per year than those with the best conditions [54]. It is important to note that this study was not randomized – that is, individuals were not assigned desks at random, and it may be that those with high status were assigned the desks with the best daylight and views. The health benefits of quality views were also suggested in the now-classic double-blind study by Roger Ulrich in 1984, where patients recovering from cholecystectomy surgery were randomly assigned rooms providing views through windows to either trees or to a brick wall. Those with views to trees recovered to levels resulting in discharge almost a day sooner, requested fewer or less potent pain medications, and had fewer notes in their files reporting discomfort or difficulty moving [55].

Case Studies: Projects Assigned with Metrics for Indoor Air Quality, Comfort, and Health

A number of recent projects led by EskewDumezRipple demonstrate that projects can be designed to deliver high indoor air quality and thermal comfort, to promote occupant physical activity, to provide daylight and views for all to achieve high energy performance – even on limited budgets.

Dalney Building, 2019

The 54,000-sf four-story Dalney Building at the Georgia Institute of Technology (Atlanta, Georgia), completed in 2019, provides 40 cfm of outdoor air per occupant through a low-friction recirculating Variable Air Volume (VAV) system with MERV 13 filtration, and achieves ASHRAE 189.1 levels of energy performance (25% better than energy code, with an Energy Use Intensity (EUI) of 37 kBtu/sf/yr). Combining low- and zero-VOC interior finishes and direct venting of areas containing printers and photocopies further helps guard indoor air quality. Thanks to a thin floorplate design, no occupant is more than 20 feet from a window providing daylight, with most provided a direct view to a newly developed landscaped hillside "Eco-Commons". Additionally, a generously proportioned stair looking out onto a landscaped courtyard connects all floors to promote physical activity. The project was designed for WELL Gold certification. At a construction cost of just $275 per square foot, the project was recognized in 2021 by the Urban Land Institute's UL10 Extra Green Building Award as one of ten exemplary projects internationally [56].

Figure 5.1 Dalney Building, 2019. Photo credit: Creative Sources Photography, Courtesy of EskewDumezRipple.

Louisiana Workers' Compensation Corporation, 2021

The 130,000-sf eight-story Louisiana Workers' Compensation Corporation (LWCC) Headquarters Building renovation (Baton Rouge, Louisiana), completed in 2021, also provides 40 cfm of outdoor air per person. It employs enthalpy recovery ventilation (ERV) wheels to lessen the energy impact of this higher ventilation rate, as part of an HVAC design combining a once-through Dedicated Outdoor Air System (DOAS) with MERV 13 filtration for ventilation with water-cooled Variable Refrigerant Flow (VRF) system for temperature control. Energy modeling used during design shows a predicted EUI of just 27 kBtu/sf/yr, just one-third the energy consumed by the building before the renovation project was undertaken, paying back the embodied carbon of the renovation project in less than a year of operational carbon savings according to lifecycle projections. The project was the first in the state to earn LEED v4 certification, and was delivered for a construction cost under $170/sf.

Figure 5.2 Louisiana Workers' Compensation Corporation, 2021. Photo Courtesy of EskewDumezRipple.

Center of Developing Entrepreneurs, 2022

The 215,000-sf eight-story Center of Developing Entrepreneurs (CODE) (Charlottesville, Virginia), completed in 2022, provides 40 cfm of outdoor air per occupant and employs enthalpy recovery ventilation (ERV) wheels to lessen the energy impact of this higher ventilation rate, as part of an HVAC design combining a once-through Dedicated Outdoor Air System (DOAS) with MERV 13 filtration for ventilation with hydronic Fan Coil Units (FCUs) for temperature control. Energy modeling used during design shows a predicted EUI of just 26 kBtu/sf/yr, a 73% reduction from the 2030 Commitment Benchmark (a survey of buildings of a range of building types performed in 2003). The project specified low- and zero-VOC interior finishes and demand-controlled ventilation for high-occupancy areas, consistent with the "Enhanced Green" strategies described by Allen. With operable windows throughout and most floors provided with extensive outdoor occupiable landscaped green roofs, the project opened two years into the COVID-19 pandemic yet was completely leased within months of opening, at a time when commercial office space occupancy rates nationally were at record highs [57]. The project also earned LEED Platinum certification and the AIA National Honor Award for design excellence.

Figure 5.3 Center of Developing Entrepreneurs, 2022. Photo credit: © Alan Karchmer, Photo Courtesy of EskewDumezRipple.

Conclusion

Buildings used to be simple, and for all their faults and discomforts, were relatively obvious to operate. The 20th century brought us huge advances in wealth, health, and comfort, and buildings that grew in complexity as their designers strove to deliver the "standard environment" – temperature, humidity, ventilation, and light levels that were thought to be good enough for the typical occupant. Yet for all their complexity, the buildings of the late 20th and early 21st centuries fall short: many occupants report that the conditions provided by buildings are uncomfortable and not conducive to health or productivity; buildings are responsible for 40 percent of climate-changing emissions; and the indoor conditions provided by buildings fostered the spread of the most severe global pandemic in a century. The challenge of the 21st century is to transform buildings from their energy-intensive, one-size-never-fits-all approach to indoor conditions to an approach that empowers occupants by providing them with easy control of their individual environment to shape the indoors to conditions that help each of us thrive. By providing all the light and air and warmth and cooling each of us needs – but only exactly where and when we need it – we can address the climate crisis while unlocking human potential.

References

[1] Fitch, James Marston. 1999. *American Building: The Environmental Forces That Shape It*. 2nd ed. New York; Oxford: Oxford University Press.

[2] Wang, Zhe, Richard de Dear, Maohui Luo, Borong Lin, Yingdong He, Ali Ghahramani, and Yingxin Zhu. 2018. "Individual Difference in Thermal Comfort: A Literature Review." *Building and Environment* 138 (June): 181–93. https://doi.org/10.1016/j.buildenv.2018.04.040.

[3] Karjalainen, S. 2012. "Thermal Comfort and Gender: A Literature Review." *Indoor Air* 22 (2): 96–109. https://doi.org/10.1111/j.1600-0668.2011.00747.x.

[4] Schaudienst, Falk, and Frank U. Vogdt. 2017. "Fanger's Model of Thermal Comfort: A Model Suitable Just for Men?" *Energy Procedia*, 11th Nordic Symposium on Building Physics, NSB2017, June 11–14, Trondheim, Norway, 132 (October): 129–34. https://doi.org/10.1016/j.egypro.2017.09.658.

[5] Chang, Tom Y., and Agne Kajackaite. 2019. "Battle for the Thermostat: Gender and the Effect of Temperature on Cognitive Performance." *PLOS One* 14 (5): e0216362. https://doi.org/10.1371/journal.pone.0216362.

[6] De Dear, Richard, and Gail Brager. 1998. "Developing an Adaptive Model of Thermal Comfort and Preference." https://escholarship.org/uc/item/4qq2p9c6.

[7] American Society of Heating, Refrigeration, and Air-Conditioning Engineers. 2004. *ASHRAE 55–2004: Thermal Environmental Conditions for Human Occupancy*. Atlanta, GA: ASHRAE. www.ashrae.org/technical-resources/bookstore/standard-55-thermal-environmental-conditions-for-human-occupancy.

[8] Van Marken Lichtenbelt, Wouter, Mark Hanssen, Hannah Pallubinsky, Boris Kingma, and Lisje Schellen. 2017. "Healthy Excursions Outside the Thermal Comfort Zone." *Building Research & Information* 45 (7): 819–27. https://doi.org/10.1080/09613218.2017.1307647.

[9] Parkinson, Thomas, and Richard de Dear. 2015. "Thermal Pleasure in Built Environments: Physiology of Alliesthesia." *Building Research & Information* 43 (3): 288–301. https://doi.org/10.1080/09613218.2015.989662.

[10] Heschong, Lisa. 1979. *Thermal Delight in Architecture*. Cambridge: MIT Press.

[11] Rea, Mark S. 2000. *The IESNA Lighting Handbook: Reference & Application*. 9th ed. New York, NY: Illuminating Engineering Society of North America. https://catalog. lib.ncsu.edu/catalog/NCSU1384465.

[12] Heschong, Lisa. 2002. "Daylighting and Human Performance." *ASHRAE Journal* 44 (6): 65.

[13] Omrany, Hossein, and Abdul Marsono. 2016. "Optimization of Building Energy Performance Through Passive Design Strategies." *British Journal of Applied Science & Technology* 13 (6): 1–16. https://doi.org/10.9734/BJAST/2016/23116.

[14] Tucker, L.M. 2021. "The Labor-Saving Kitchen: Sources for Designs of the Architects' Small Home Service Bureau." *Enquiry: The ARCC Journal for Architectural Research*, June. http://arcc-repository.org/index.php/arccjournal/article/view/208.

[15] Furchgott, Roy. 2016. "Controlling Steam Radiators." *The New York Times*, March 11, sec. Real Estate. www.nytimes.com/2016/03/13/realestate/radiators-steam-heat-temperature-control.html.

[16] Hobday, Richard A., and John W. Cason. 2009. "The Open-Air Treatment of Pandemic Influenza." *American Journal of Public Health* 99 (S2): S236–42. https://doi. org/10.2105/AJPH.2008.134627.

[17] Holohan, Dan. 1992. *The Lost Art of Steam Heating*. Bethpage, NY: Dan Holohan Associates, Incorporated.

[18] Nelissen, Sander, and Mariël Polman. 2012. "Duikers' Open Air School: Re-Use or Contin-Use?" *Docomomo Journal* 47 (December): 34–41. https://doi. org/10.52200/47.A.4LYGANH7.

[19] Moe, Kiel. 2010. *Thermally Active Surfaces in Architecture*. Princeton: Princeton Architectural Press.

[20] Allen, Joseph G., Piers MacNaughton, Usha Satish, Suresh Santanam, Jose Vallarino, and John D. Spengler. 2016. "Associations of Cognitive Function Scores with Carbon Dioxide, Ventilation, and Volatile Organic Compound Exposures in Office Workers: A Controlled Exposure Study of Green and Conventional Office Environments." *Environmental Health Perspectives* 124 (6): 805–12. https://doi. org/10.1289/ehp.1510037.

[21] Marsh, Allison. 2019. *The Factory: A Social History of Work and Technology*. History of Human Spaces. Santa Barbara, CA: Greenwood, an Imprint of ABC-CLIO, LLC. https://catalog.lib.ncsu.edu/catalog/NCSU4551821.

[22] US EPA. 2014. "A Brief Guide to Mold, Moisture and Your Home." *Overviews and Factsheets*, August 13. www.epa.gov/mold/brief-guide-mold-moisture-and-your-home.

[23] Fisk, William J., Olli Seppanen, David Faulkner, and Joe Huang. 2004. *Economic Benefits of an Economizer System: Energy Savings and Reduced Sick Leave*. LBNL 54475. Berkeley, CA: Lawrence Berkeley National Lab (LBNL). www.osti.gov/ biblio/821457.

[24] Ackerly, Katie, and Gail Brager. 2012. "Human Behavior Meets Building Intelligence: How Occupants Respond to 'Open Window' Signals." August. https://escholarship. org/uc/item/0835d5w4.

[25] WELL. n.d. "WELL Standard V2." Accessed December 5, 2022. https://v2.wellcertified. com/en/performance-rating/overview.

[26] Brager, Gail, Gwelen Paliaga, and Richard de Dear. 2004. "Operable Windows, Personal Control and Occupant Comfort." https://escholarship.org/uc/item/4x57v1pf.

[27] Rijal, H. B., P. Tuohy, M. A. Humphreys, J. F. Nicol, A. Samuel, and J. Clarke. 2007. "Using Results from Field Surveys to Predict the Effect of Open Windows on Thermal Comfort and Energy Use in Buildings – Getting Them Right." *Energy and Buildings* 39 (7): 823–36. https://doi.org/10.1016/j.enbuild.2007.02.003.

[28] Wang, Liping, and Steve Greenberg. 2015. "Window Operation and Impacts on Building Energy Consumption." *Energy and Buildings* 92 (April): 313–21. https://doi.org/10.1016/j.enbuild.2015.01.060.

[29] Appleton, Jay. 1975. *The Experience of Landscape*. London; New York: Wiley.

[30] Nakano, Junta, and Shin-ichi Tanabe. 2004. "Thermal Comfort and Adaptation in Semi-Outdoor Environments." *ASHRAE Transactions* 110: 543–53.

[31] Messier, K. P., L. G. Tidwell, C. C. Ghetu, D. Rohlman, R. P. Scott, L. M. Bramer, H. M. Dixon, K. M. Waters, and K. A. Anderson. 2019. "Indoor Versus Outdoor Air Quality During Wildfires." *Environmental Science & Technology Letters* 6 (12): 696–701. https://doi.org/10.1021/acs.estlett.9b00599.

[32] Liang, Yutong, Deep Sengupta, Mark J. Campmier, David M. Lunderberg, Joshua S. Apte, and Allen H. Goldstein. 2021. "Wildfire Smoke Impacts on Indoor Air Quality Assessed Using Crowdsourced Data in California." *Proceedings of the National Academy of Sciences* 118 (36): e2106478118. https://doi.org/10.1073/pnas.2106478118.

[33] American Society of Heating and Ventilating Engineers. 2017. *ASHRAE 52.2–2017: Method of Testing General Ventilation Air-Cleaning Devices for Removal Efficiency by Particle Size*. ANSI/ASHRAE Standard, 1041–2336; 52.2–2012. Atlanta, GA: ASHRAE. www.ashrae.org/File%20Library/Technical%20Resources/COVID-19/52_2_2017_COVID-19_20200401.pdf.

[34] Azimi, Parham, and Brent Stephens. 2013. "HVAC Filtration for Controlling Infectious Airborne Disease Transmission in Indoor Environments: Predicting Risk Reductions and Operational Costs." *Building and Environment* 70 (December): 150.

[35] Morawska, Lidia, Julian W. Tang, William Bahnfleth, Philomena M. Bluyssen, Atze Boerstra, Giorgio Buonanno, Junji Cao, et al. 2020. "How Can Airborne Transmission of COVID-19 Indoors Be Minimised?" *Environment International* 142 (September): 105832. https://doi.org/10.1016/j.envint.2020.105832.

[36] Dal Porto, Rachael, Monet N. Kunz, Theresa Pistochini, Richard L. Corsi, and Christopher D. Cappa. 2022. "Characterizing the Performance of a Do-It-Yourself (DIY) Box Fan Air Filter." *Aerosol Science and Technology* 56 (6): 564–72. https://doi.org/10.1080/02786826.2022.2054674.

[37] Center for Disease Control and Prevention. 2023. "Ventilation in Buildings." Centers for Disease Control and Prevention, May 12. www.cdc.gov/coronavirus/2019-ncov/community/ventilation.html.

[38] American Society of Heating and Ventilating Engineers. 2023. "ASHRAE Completes Draft of First-Ever Pathogen Mitigation Standard." May 15. www.ashrae.org/about/news/2023/ashrae-completes-draft-of-first-ever-pathogen-mitigation-standard.

[39] U.S. Green Building Council. n.d. "Design for Indoor Air Quality and Infection Control." Accessed May 22, 2023. www.usgbc.org/credits/safety-first-155-v4.1?view=language&return=/credits/New%20Construction/v4.1.

[40] Hill, Leonard, and Martin Flack. 1908. "The Effect of Excess of Carbon Dioxide and of Want of Oxygen upon the Respiration and the Circulation." *The Journal of Physiology* 37 (2): 77–111.

[41] National Institute for Occupational Safety and Health. 1994. "Carbon Dioxide." May. www.cdc.gov/niosh/idlh/124389.html.

[42] NASA. n.d. "Apollo 13 Flight Journal – Day 4, Part 4: Building the CO2 Adapter." https://history.nasa.gov/afj/ap13fj/15day4-mailbox.html.

[43] Persily, Andrew. 2022. "Development and Application of an Indoor Carbon Dioxide Metric." *Indoor Air* 32 (7): e13059. https://doi.org/10.1111/ina.13059.

[44] Persily, Andrew. 1997. "Evaluating Building IAQ and Ventilation with Indoor Carbon Dioxide." *ASHRAE Transactions* 103, Part 2, Proceedings of the ASHRAE Summer Meeting. www.aivc.org/resource/evaluating-building-iaq-and-ventilation-indoor-carbon-dioxide.

[45] American Society of Heating and Ventilating Engineers. 2010. *ASHRAE 62.2–2010: Ventilation for Acceptable Indoor Air Quality*. ANSI/ASHRAE Standard. Atlanta, GA: ASHRAE.

[46] Satish, Usha, Mark J. Mendell, Krishnamurthy Shekhar, Toshifumi Hotchi, Douglas Sullivan, Siegfried Streufert, and William J. Fisk. 2012. "Is CO2 an Indoor Pollutant? Direct Effects of Low-to-Moderate CO2 Concentrations on Human Decision-Making Performance." *Environmental Health Perspectives* 120 (12): 1671–77. https://doi.org/10.1289/ehp.1104789.

[47] MacNaughton, Piers, James Pegues, Usha Satish, Suresh Santanam, John Spengler, and Joseph Allen. 2015. "Economic, Environmental and Health Implications of Enhanced Ventilation in Office Buildings." *International Journal of Environmental Research and Public Health* 12 (11): 14709–22. https://doi.org/10.3390/ijerph121114709.

[48] Laurent, Jose Guillermo Cedeño, Piers MacNaughton, Emily Jones, Anna S. Young, Maya Bliss, Skye Flanigan, Jose Vallarino, Ling Jyh Chen, Xiaodong Cao, and Joseph G. Allen. 2021. "Associations Between Acute Exposures to PM2.5 and Carbon Dioxide Indoors and Cognitive Function in Office Workers: A Multicountry Longitudinal Prospective Observational Study." *Environmental Research Letters* 16 (9): 094047. https://doi.org/10.1088/1748-9326/ac1bd8.

[49] American Society of Heating and Ventilating Engineers. 2022. "ASHRAE Position Document on Indoor Carbon Dioxide." www.ashrae.org/file%20library/about/position%20documents/pd_indoorcarbondioxide_2022.pdf.

[50] Milton, Donald K., P. Mark Glencross, and Michael D. Walters. 2000. "Risk of Sick Leave Associated with Outdoor Air Supply Rate, Humidification, and Occupant Complaints: Sick Leave and Building Ventilation." *Indoor Air* 10 (4): 212–21. https://doi.org/10.1034/j.1600-0668.2000.010004212.x.

[51] Heschong, Lisa. 1999. "Daylighting in Schools: An Investigation into the Relationship Between Daylighting and Human Performance. Detailed Report." https://eric.ed.gov/?id=ED444337.

[52] Heschong Mahone Group, Inc. 2003. *Windows and Offices: A Study of Office Worker Performance and the Indoor Environment*. Technical Report. Fair Oaks, CA: California Energy Commission.

[53] Heschong, Lisa. 2003. *Daylight and Retail Sales*. Gold River, CA: Heschong Mahone Group. https://newbuildings.org/wp-content/uploads/2015/11/A-5_Daylgt_Retail_2.3.71.pdf.

[54] Elzeyadi, Ihab. 2011. "Daylighting-Bias and Biophilia: Quantifying the Impact of Daylighting on Occupant Health." www.researchgate.net/profile/Ihab-Elzeyadi/publication/344361245_Daylighting-Bias_and_Biophilia_Quantifying_the_Impacts_of_Daylighting_and_Views_on_Occupants_Health/links/5f6c4023458515b7cf497bc6/

Daylighting-Bias-and-Biophilia-Quantifying-the-Impacts-of-Daylighting-and-Views-on-Occupants-Health.pdf.

[55] Ulrich, Roger S. 1984. "View Through a Window May Influence Recovery from Surgery." *Science* 224 (4647): 420–1. https://doi.org/10.1126/science.6143402.

[56] Nyren, Ron. 2021. "UL10: Extra-Green Buildings." *Urban Land Magazine*, March 29. https://urbanland.uli.org/planning-design/ul10-extra-green-buildings/.

[57] Irwin-Hunt, Alex. 2023. "Out of Office: US Vacancy Rates Hit Record High." *FDi Intelligence*, April 5. www.fdiintelligence.com/content/data-trends/out-of-office-us-vacancy-rates-hit-record-high-82348.

6 Respiratory Equity

Ulysses Sean Vance

The basic human needs and rights we hold to be inalienable are directly connected to space. They are spatial in that everything humans require for survival – air, water, food, and shelter – is tied to an ecology greater than the space of our individual bodies alone [1]. As such, all these needs together define our engagement, whether as individuals or civic societies, and in some way frame our environmental reliance. It is due to this environmental reliance, along with the urgency that comes from sharing limited resources, that society is constantly being redefined [2]. Architects and architecture, as active players in redefining the ecologies of human settlement, lay the foundations for spatial tensions in society. The fragility of modern society is often related to our misconceptions of the roles and rules that govern responsibilities, and these misconceptions are exacerbated by differences in culture. However, the spatiality of humanity and human needs remains constant.

Respecting human rights means that everyone has the right to life, liberty, and security of person [3]. Equally, the contemporary concept of architecture, which for the sake of this argument involves an ecological and environmental approach to wellness, is the spatial manifestation of the right to life and the security of the person. Architecture's involvement in helping define liberty, which for the sake of this argument is here defined as being parallel with freedom, as in "choosing your responsibility, not having no responsibilities," [4] began shortly after World War II. It was at this time that the new intergovernmental organization, the United Nations, formed a council to address the "commissioning in economic and social fields" of means "for the promotion of human rights." [5] This Economic and Social Council (ECOSOC), which at the time was chaired by Eleanor Roosevelt, was formed to examine the specifics of human rights, and it ultimately produced the "Universal Declaration of Human Rights." Under the charter of the United Nations, these human rights are defined spatially through the acceptance or reluctance of certain countries to participate. The intention of the commission at that time was to restore to individuals the human dignity lost during war and to provide access to systems that supported their recovery. Those UN-defined human rights have as much relevance today, as they include equal

DOI: 10.1201/9781003398714-7

and equitable rights to air, water, adequate food, and shelter. The initial intentions of the UN commissions still serve as guidelines for designing a sense of personal security into places of habitation.

Needs and Rights: Equity in Design

Historically, tensions have arisen in the application of human rights to architecture because of individual or group perceptions about who has the right to occupy particular spaces [6,7]. The practice of reserving some spaces for certain people was vehemently challenged in the United States as far back as the Civil Rights Act of 1866, which came to prominence in the Civil Rights Movement between 1942 and 1968, and came to the fore again in the movement that led to the Americans with Disabilities Act (1990). From an architectural standpoint, rights are critical to determining equity, especially as they relate to the accessibility of a space and its ability to maintain service for needs beyond that of an individual alone. In preparing to address human rights and needs directly in architecture, the designer's work of isolating the functionality of body systems helps establish principal measures to preserve and enhance human dignity, comfort, and trust. From such work, a framework for designing toward more equitable applications can be established.

As there are many plausible approaches to this endeavor of equitable design, the selection of the most important criteria focuses first on survival and then on establishing air, water, food, and shelter as the prime imperatives of establishing equity. In working to address the promotion of human needs and subsequently human rights in design, it is helpful to place the new criteria into a framework of assessments and strategies, moving from the implementation of theories into a design concept and applying the most acute observations to the engagement of needs and rights at the place of intervention. And while identifying these criteria relative to human needs and human rights aids in assessing equivalent survival capacity, they do not guarantee a design will be immediately equitable when providing services. There are many elements of equity that are only implementable through the services themselves. In these cases, the architecture should not detract from the encounter, even when the failings of the encounter relate to exclusionary boundaries in the environment.

Air as a Human Right (Breathing)

Traditionally, architectural design responds to the properties of a condition and, as with the direct approach of sheltering, establishes safeguards for the occupants when the property itself has less-desirable features. As the number of persons increases in a particular space, its capacity to safeguard the occupants against negative environmental circumstances diminishes. From a space planning perspective, the typical architectural design response is to first prioritize maintaining security for activities against the prevailing issues and then attend to their adjacencies based on assessment of the conditions, while reinforcing the boundaries between the activities [8]. These reassessments form hierarchies in the use of the space, and it is these hierarchies that create inequities that later become disparities after prolonged neglect [9]. Inverting these practices would situate the disparity as the top priority when establishing solutions for balancing against the inequities; however, this would not directly contribute to the design of a good solution. The result would be a design safeguarding against the disparity but leaving in place the negative condition(s), thus compounding the problem. A more appropriate approach would be to address the entirety of the challenge simultaneously, taking advantage of integrating approaches that enable all phases of the design to contribute to the response.

In addressing the challenges of identifying a framework of procedures on how equity can inform architectural design from inception, through intervention, and into post-occupancy evaluation, it would actually be easier to address the first of these frameworks in isolation, as the discussion of such an approach would be too broad for the context of this narrative. When focusing on the specifics of health equity, for example, respiratory rights and the value in being able to access breathable air, it is important to consider the social aspects of the conditions. In the case of respiratory equity, the conditions of open space, open air, and how the perceptions of familiarity, comfort, and trust over time in a space contribute to perceived equitability in care [10]. **Respiratory equity is established not only in the physical act of breathing but also in the ability to maintain access to an expanse of air that would at the same time contribute positively to the mental attributes of breathing. Thus, respiratory equity defines and is defined by prolonged access to open space.**

"Breathing" is a general term that describes the life-sustaining effort of moving air in and out of the body, with the action's mechanical effectiveness determined by the capacity of the lungs and the rate of inhalation and exhalation by the diaphragm, the muscle responsible for filling and emptying the lungs. In an architectural sense, the breathability of a space can be determined by a variety of factors, including the amount of internal open space connected to the outdoors, the quantity of building elements that act as apertures connecting interior spaces to outdoor conditions, and the capacity of the mechanical system to provide a preferred level of exchange between interior and exterior air [11]. The capacity for building elements to prioritize the relationship between interior and exterior conditions, as in the presence of a porch, promenade, or breezeway, can be influential in providing adequate air movement for each occupant's optimal breathing. Each of these elements is important in providing measurable approaches to connecting indoor and outdoor space when designing building enclosures, because they determine the respiratory capacity of the building for occupants. It is equally valuable to determine whether the conditions are temporarily or permanently open, and having this quantitative and qualitative measure of breathability makes it possible to address the health disparities that form in the absence of architectural elements working toward equitable solutions.

The COVID-19 pandemic has heightened awareness of the need to design for greater air exchange capacities. Attention to the frequency of airborne pathogens inside buildings is shifting awareness from prioritizing long-term energy efficiency to a focus on airtight building envelopes that include broader considerations of outdoor air access [12]. With this urgent need for a shift in focus, design must offer new ways to allow buildings and their occupants to inhale and exhale air safely from their respective bodies. For much of the modern period in architecture, concerns about breathing were primarily industrial and confined to controlling poisons through acts and regulatory measures. These and other oversimplifications associated with regulating air at the industrial level ignore breathing inside buildings and hinder efforts to increase the connectivity between interior and exterior spaces. As a result, the primarily policy-driven approach to air has proven to be a cautionary tale, as it has focused almost entirely on occupational hazards, resulting in practices incorporating the mechanics and technology of sensing the hazards over passive building elements such as porches, courtyards, and breezeways to reduce concentration of hazards and thus had little impact on improving the indoor air of living spaces [13].

The COVID-19 pandemic and the resulting closure of offices, schools, and dining and entertainment venues suddenly shifted focus on air quality from work settings to domestic ones. As the level of social interaction increased in domestic spaces, the mirroring of standards in this dynamic shift from workplaces to domestic spaces rather tentatively brings to light the challenges of safe breathing at home. Considering the social and behavioral differences between domestic and civic spaces is also important to improve air exchange within confined spaces across housing typologies. It is particularly important in dense urban housing, as these building types often have a legacy of disparities and their impact on the low-income communities is considerably greater, since these developments are often situated downwind of factories or farms that emit hazardous fumes, whether offensive to the senses or completely unseen and unscented [14].

Economically disadvantaged populations are more likely to live in proximity to hazardous activities worldwide. In the United States the disregard for such a predicament has been most impactful on low-income communities, which are often relocated to hazardous areas systematically through the devaluation of property and limited availability of social, or public, housing [15]. Whether directly or indirectly, programming for passive ventilation in enclosed spaces certainly appears more equitable than physically limiting ventilation to mechanical systems. Practices that impact the building envelope's respiratory character and force a reliance on heavily industrialized mechanical systems to provide fresh air devalue shared spaces because of improper ventilation. Because low-income communities are more likely to live in shared spaces, design that fails to introduce fresh air is inherently endangering health in a time when pathogenic disease heavily impacts those living in spaces subject to recirculated, unfiltered indoor air.

An Architectural Approach

In defining respiratory equity in architecture, establishing the building as a device for perpetuating equitable interpretations on breathability is about augmenting performance metrics with those focused on the healthy qualities of space. By focusing on the feasibility of architecture to perform certain tasks associated with increasing airflow, facilitating air exchanges, and filtering pollutants – tenets of breathability – the intentions in the applications are to directly assist communities where neither sustainability nor healthfulness have been applied. And by reframing healthfulness as the level of wellness in a space, breathability is introduced as a necessary metric for building design.

The American Thoracic Society interprets respiratory health equity as "the attainment of the highest level of respiratory health for all people." [16] Its position on the framework of respiratory equity as it relates to health equity is from a medical standpoint and focuses on "valuing everyone equally, implementing and maintaining focused societal efforts to address avoidable inequalities and historical and contemporary injustices, and eliminating health care disparities." [17] In terms of architectural interventions, the characteristics for addressing the broadness of health equity and the specifics of respiratory equity are primarily focused on the criteria for incorporating systems for air movement into the interconnectedness of activities, materials, assemblies, and procedures based on the use and utility of a space. And while the desirability of "attainment of the highest level of respiratory health for all people" [17] in inhabited spaces is seemingly obvious, each architectural element's involvement in fulfilling these criteria, in part or in whole, inevitably must address the boundaries between inside and outside, which is the area that offers the most significant potential for improvement.

While there are many ways to approach the challenges of respiratory equity through architecture, the intention of this narrative is to discuss the feasibility of establishing an attitude that allows architects to think about breathability in a spatial sense, thus arguing for the significant role of architecture in health and wellness. These practices have been in use across multiple courses taught by the author that focus on associating architectural concepts with medical terminology. These associations reframe the challenges of designing ecologies for bettering wellness from what makes for a "sick building" toward proactively incorporating an understanding of body systems and their corresponding needs as part of the environment, from the scale of an object in a room to that of a building in a city, and aspirationally even greater global contexts.

A tacit understanding of the mechanics of ventilation is beneficial to designing buildings as devices that contribute to healthy airflow. For example, working with architecture students to associate medical terminology with respiratory care for the body – such as the medical term respiratory rate (the number of breaths a person takes in a minute) with the broader term ventilation (the movement of air through the conducting passages between the atmosphere and the lungs) – helps their comprehension of the design challenges by personalizing pulmonary ventilation and then correlating the issues and criteria for improving emissions in building design. The synergizing of vocabularies also works to establish a baseline for broadening the design team, incorporating medical specialties into the practices of the design studio and reimaging the work of design from individual thought to a more collaborative practice and approach.

From these introspective and collaborative approaches, the particulars of designing space by means associated with respiratory equity go from comprehending health as an individual assessment of the challenges to more civic-mindedness about wellness. Opening up the design challenge to broader comprehensions of the issues around respiratory equity is beneficial to articulating safety in service to breathing at any scale. Strategies for doing so include establishing scalable references to personal breathing as the project increases from building element to room to building, as well as remaining in contact with individual and civic references for measuring the success or failure of concepts across changes in settings and capacity. The primary distinction is to connect design strategies and design evidence for ventilation at each scale of the conceptual thinking around the building as a breathing device, by way of element, room, and building, actively incorporating testing concepts immediate to the body along the way, a concept that was addressed more broadly in a previous chapter (Harmon). In addressing health along this escalating scale, prototyping a building as a breathing device reframes the act of breathing in the design studio among colleagues, as each breath informs the understanding and application of these everyday challenges, albeit at a more civic scale, while informing the narrative of breathability against various environmental challenges.

On Architectural Characteristics in Light of COVID-19

Learning to apply these new methods in the design studio involves both awareness of novel approaches to the challenges and evidence-based solutions that focus on respiratory equity. Well past the COVID spikes of 2020 and 2021, the global death toll passed the six million mark in 2022 and global case counts exceeded 500 million [18]. Accompanying the stories of the lives lost were the enormous challenges for the attending health services struggling to care for the large number of patients in the facility where they were diagnosed, while many others with less severe symptoms returned home. The caveat or assumption was that the latter group of patients would self-isolate until their condition improved. Unfortunately, for many communities in both the low- and middle-income brackets, the return home induced an escalation in disease transmission, as their home environment was often a shared space within a multifamily structure [19]. Additionally, the lack of airtight separation for the building enclosure and inadequate spacing between units bring into question the level of feasible maintenance and its use in preventing airborne disease transmission. Accompanying these challenges is a "missing middle" of constructs for patients who cannot self-isolate and must therefore spend the duration of their "isolation" and recuperation period in congregate healthcare settings [20].

As can easily be imagined, the events related to the COVID-19 pandemic have had a tremendous impact on the work of teaching architecture students about health equity and respiratory equity. Many of the students were experiencing firsthand the challenges in the design problem. Challenges that in years prior to the pandemic were often perceived only at a distance from the discourse were now directly impacting everyone's lives. For example, as the medical industry learned in 2020 when the COVID pandemic caused an abrupt shift to telemedicine services, not all consumers had access to smartphones or computers with high-speed internet access [21]. These challenges were broadly detrimental, affecting almost every industry, as online service challenges proved catastrophic to a variety of businesses. Similarly, when K–12 schools shifted to online learning, low-income families were disproportionately impacted by a lack of technology resources; this negative impact will outlast the COVID-19 pandemic [22]. Moreover, when families did have access to technologies such as computers or smartphones, the applications were often subscription-based services, the costs of which were and may still be prohibitive for some and thus an impediment to widespread utilization.

Additionally, the challenges of the technology gap continue to isolate low-income communities, as they must continuously prioritize other, more immediate personal expenditures and therefore not be able to address the larger issue of their poor housing conditions. This self-protective pattern of prioritizing immediate over long-range needs is precisely why "expert" advice that suggests displacing the person or family from a place of intended safety in the home to a place of assumed safety, as in by *escaping to a park for some fresh air*, is problematic. Such advice also removes the "inalienability" aspect of human rights, in that it denies a person the ability to experience the right to live in the protection of their home. Whether the challenge of applying an equitable prototype of housing in service to health at this scale is conceivable or not, the benefits of addressing the toxicity of confined spaces and the environmental challenges of respiratory equity validate the establishment of buildings as a protective environment created from a framework of passive ventilation practices. Otherwise, the disconnect between expensive ventilation systems for financially constricted communities will continue to exacerbate the issues in affordable housing related to the COVID-19 pandemic.

Forecasting Plausibility

In determining the feasibility of responding to these and other complex situations, including the distinct constructs that form around people as they breathe in and out, it is imperative to the study of health equity to track and evaluate what is unique to people's circumstances when addressing ways to improve their condition. In isolating each of the factors affecting respiratory issues at this stage of the work, it is necessary to examine how each service operates separately based on the ventilation of a volume of space. By isolating the characteristics in each room and identifying a framework of procedures for domestic and institutional environments, we can evaluate the factors of inviting clean air in and exhausting bad air, both of which may be improved with the filtering of pollutants. For confined spaces, it is critical to determine if the challenge is defined by the need for the area to be completely isolated or if the isolation results from spaces surrounding its location have high priority in the arrangement. Changing the siting in the preliminary planning and programming stage to allow spaces to ventilate immediately to the outside will be easier than relying on a mechanical system to evacuate contaminated air.

By studying the feasibility of transitioning specific medical interventions into a home healthcare model for low-income communities, we may be able to bridge the medical services gaps between institutional health facilities and domestic settings. However, telemedicine elements have limited utility without the ability to relay information from the home environment to the doctor. The use of medical share-points and personal devices helps, but concerns persist regarding the environmental strategies for maintaining adequate protections once a medical problem is diagnosed. The intervention of shared medically oriented spaces makes health services more accessible and presents alternative means of informing communities about airborne transmission of disease in urban environments. Formal means of information dissemination include extending the conversation on intelligent spaces as devices, knowing when indoor environments are hazardous (whether in civic, domestic, or institutional settings), and applying these practices actively in the field through passive building elements that convey this understanding in an intuitive manner.

Respiratory Equity in Action

By chronicling air movement across more extensive territories, a design focused on integrating human-centered design goals would include the ability of persons to adjust or actively select places in which to conduct activities based on environmental preferences. The most prevalent application of such an arrangement would be from the food-services industry, which during the pandemic undertook efforts to continue serving customers in urban environments. Not all of these interventions were good, but in some cases the interventions assisted in improving a city's resilience and ecological characteristics in areas where challenges such as urban heat gain or concentrations of pollutants are causing the most damage. One example was the reassessment of restaurant space, which included assessing the types of conflict between staff and customers and the general spatial conditions which, if served by an additional layer of space, would provide greater access to outdoor air. For one intervention, the new measures included utilizing vacant spaces for commercial squatting to mitigate the proximity of negative environmental conditions. This approach was based on evaluating conditions for the effectiveness of the outdoors as an integral building element that maximize protections while facilitating preferences for particular experiences. The architect's participation involved establishing the physical properties of mobility for the kitchen against a changing field of outdoor elements that allow for an intriguing accumulation of understandings on how environmental adjacency can be instrumental in enhancing engagement. These measures take advantage of the conditions in place of forecasting for systems and represent the capacity to see and make changes to the environment, which directly influences architectural elements and urban strategies for adapting to the challenges of working during the COVID-19 pandemic.

The COVID experience and field observations from restaurants, schools, and medical spaces have motivated new trajectories of design that underscore the health benefits of adding amenities such as outdoor porches and interior breezeways to more domestic spaces, such as multifamily housing. The ability to ventilate confined spaces by incorporating porches into the design articulates the value of open-air systems in decreasing the concentration of transmissible particulates in the air. In addition, this knowledge is beneficial in helping occupants of low-income housing rearrange living conditions to encourage natural air movement, which is substantially dependent on flexibility in the building design, spacing between units, availability of cross-ventilating windows, and the internal arrangement of rooms.

Musings about Rights and Breathing

In designing urban spaces and building elements to reflect an agenda focused on respiratory equity, the macro studies of healthcare, with their vocabularies and settings, can inform more minor applications in the domestic spaces where personal well-being begins and evolves. As the transmission of airborne disease in nonmedical spaces has continued and even intensified with the mutating nature of the COVID-19 virus, it is widely speculated that biosafety references will continue to influence space-planning through the design of temporary and permanent care facilities. As a result, this discussion also establishes that further study of the criteria for querying the spatial arrangements of social elements, furnishings, and equipment is needed to determine the best engagement between people, whether at home or in medical spaces, and between care provider and patient. These further considerations include reducing the cross-contamination of non-adjoining residences and being mindful of apertures when arranging furnishing, equipment, and barriers in response to airflow. Additional measures would include the integration of a synthesized approach combining environmental understanding from architecture, interior architecture, and landscape architecture toward "open space" in an existing building; evaluating the feasibility of adding green space to the programming; and supporting air stagnation mitigation by tenants themselves. The most critical aspect is identifying the benefits of these applications in restoring the personal connection and dignity that many feel has been lost as a result of the pandemic experience.

It must be acknowledged that human-centered design thinking, particularly with regard to people's need for and right to medical services at appropriate facilities, as well as similar thinking applied in contesting the isolation of medical areas in urban settings, benefits the broader community in familiar ways. As designers applying these concepts, we need to be doing more than predicting and programming; instead, **we must work to close the gap between services and the communities they serve.** If medical care facilities that are embedded in communities can render health services directly to an audience through both medical and nonmedical platforms, we need to be examining how other facility types and civic centers can facilitate both medical and nonmedical needs. As both an active zone and presence, the airspace of healthcare offers an intriguing set of arguments for the importance of architecture in terms of health for people – be they temporary visitors or permanent residents – who encounter buildings. Central to the moments processed through these exploratory works of architecture is the framing of hospitality and health as driving the central argument for both respiratory equity and the broader implications of health equity as a built-in tenet of airspace in building activities, materials, assemblies, and systems. We must test the criticality of a healthy environment as a protagonist of universal human rights.

References

[1] National Aeronautics and Space Administration (NASA). 2006. "Exploration: Then and Now – Survival! Lesson." *STEM Resources and Opportunities, Brian Dunbar*. www.nasa.gov/stem-ed-resources/jamestown-survival.html.

[2] Durkheim, Emile. 1984. *The Division of Labor in Society*. Basingstoke: Macmillan.

[3] United Nations. 1948. *Universal Declaration of Human Rights*. New York: United Nations, Department of Public Information.

[4] The Nobel Prize [@NobelPrize]. 2020. "Freedom Is Choosing Your Responsibility." Tweet. *Twitter*. https://twitter.com/NobelPrize/status/1229836743836389376.

[5] United Nations. n.d. "Economic and Social Council." Accessed August 3, 2022. www.un.org/ecosoc/en/content/about-us.

[6] O'Brien, William E. 2015. *Landscapes of Exclusion: State Parks and Jim Crow in the American South*. Amherst: University of Massachusetts Press.

[7] Cheng, Irene, Charles L. Davis, and Mabel O. Wilson. 2020. *Race and Modern Architecture: A Critical History from the Enlightenment to the Present*. Pittsburgh: University of Pittsburgh Press.

[8] Lang, Jon T. 1987. *Creating Architectural Theory: The Role of the Behavioral Sciences in Environmental Design*. New York: Van Nostrand Reinhold Company.

[9] Engineering National Academies of Sciences, Health and Medicine Division, Board on Population Health and Public Health Practice, Committee on Community-Based

Solutions to Promote Health Equity in the United States, Alina Baciu, Yamrot Negussie, Amy Geller, and James N. Weinstein. 2017. "The Root Causes of Health Inequity." In *Communities in Action: Pathways to Health Equity*. Washington, DC: National Academies Press. www.ncbi.nlm.nih.gov/books/NBK425845/.

[10] Bowlby, Sophie. 2012. "Recognising the Time-Space Dimensions of Care: Caring-scapes and Carescapes." *Environment and Planning* 44 (9): 2101–18.

[11] Heschong, Lisa. 1979. *Thermal Delight in Architecture*. Cambridge: MIT Press.

[12] World Health Organization. n.d. "Report: COVID-19 Slows Progress Towards Universal Energy Access." Accessed August 3, 2022. www.who.int/news/item/01-06-2022-report-covid-19-slows-progress-towards-universal-energy-access.

[13] Gaffney, Adrienne. 2019. "The Rise and Fall of the Front Porch." *Wall Street Journal*, September 11, sec. Real Estate. www.wsj.com/articles/the-rise-and-fall-of-the-front-porch-11568206837.

[14] Mohai, Paul, and Robin Saha. 2015. "Which Came First, People or Pollution? Assessing the Disparate Siting and Post-Siting Demographic Change Hypotheses of Environmental Injustice." *Environmental Research Letters* 10 (11): 115008. https://doi.org/10.1088/1748-9326/10/11/115008.

[15] Bullard, Robert D. 2000. *Dumping In Dixie: Race, Class, and Environmental Quality*. Boulder, CO: Westview Press.

[16] American Thoracic Society and Health Equity Committee. n.d. "Health Equity." Accessed August 3, 2022. www.thoracic.org/about/health-equality/.

[17] Celedón, Juan C., Esteban G. Burchard, Dean Schraufnagel, Carlos Castillo-Salgado, Marc Schenker, John Balmes, Enid Neptune, et al. 2017. "An American Thoracic Society/National Heart, Lung, and Blood Institute Workshop Report: Addressing Respiratory Health Equality in the United States." *Annals of the American Thoracic Society* 14 (5): 814–26. https://doi.org/10.1513/AnnalsATS.201702-167WS.

[18] Ritchie, Hannah, and Max Roser. 2017. "Air Pollution." *Our World in Data*, April. https://ourworldindata.org/coronavirus.

[19] Lei, Hao, Xiaolin Xu, Shenglan Xiao, Xifeng Wu, and Yuelong Shu. 2020. "Household Transmission of COVID-19-a Systematic Review and Meta-Analysis." *Journal of Infection* 81 (6): 979–97. https://doi.org/10.1016/j.jinf.2020.08.033.

[20] Akaishi, Tetsuya, Shigeki Kushimoto, Yukio Katori, Shigeo Kure, Kaoru Igarashi, Shin Takayama, Michiaki Abe, et al. 2021. "COVID-19 Transmission in Group Living Environments and Households." *Scientific Reports* 11 (1): 11616. https://doi.org/10.1038/s41598-021-91220-4.

[21] Omboni, Stefano, Raj S. Padwal, Tourkiah Alessa, Béla Benczúr, Beverly B. Green, Ilona Hubbard, Kazuomi Kario, et al. 2022. "The Worldwide Impact of Telemedicine During COVID-19: Current Evidence and Recommendations for the Future." *Connected Health* 1 (January): 7–35. https://doi.org/10.20517/ch.2021.03.

[22] Qian, Alexander S., Melody K. Schiaffino, Vinit Nalawade, Lara Aziz, Fernanda V. Pacheco, Bao Nguyen, Peter Vu, Sandip P. Patel, Maria Elena Martinez, and James D. Murphy. 2022. "Disparities in Telemedicine During COVID-19." *Cancer Medicine* 11 (4): 1192–201. https://doi.org/10.1002/cam4.4518.

7 Transcript of Panel Discussion

Inside/Out Symposium

March 31, 2022. Charlotte, North Carolina

This section represents the panel discussion from the Inside/Out Symposium in Charlotte, North Carolina, which brought together the authors from the previous four chapters, Sarah Haines, Marcel Harmon, Z Smith, and Ulysses Sean Vance, moderated by Elizabeth (Liz) McCormick. The event took place on March 31, 2022, as we were emerging from the worst of the COVID-19 pandemic. The conversation was set in a covered, semi-conditioned pavilion at the NoDa Brewing Company to discuss the state of indoor space. Despite heavy rains during the first part of the day, we were able to open up the pavilion walls to have this conversation with direct access to the outdoors. This transcript has been edited for clarity and brevity, though special attention has been paid to preserving the spirit of what was said by the panelists, while streamlining the dialogue and removing any extraneous or repetitive content.

Liz McCormick: Enhanced ventilation and energy efficiency can be at odds. We're in a non-airtight space now and I wonder when you're using airtightness as an efficiency measure, how do you balance the energy loss through increased ventilation versus the productivity gain?

Z Smith: The slogan we've all been taught is "build tight, ventilate right", and I think that the challenge is that we're taught to make the *building* comfortable rather than the *people*. The goal is to provide all the ventilation you could dream about . . . but only where and when you need it. We don't care about ventilating the building when no one's in it. ASHRAE standards are actually built around the idea that you still ventilate a building even when it's empty because the architects and interior designers specified some really nasty stuff.

Liz McCormick: My problem with airtight buildings is that you *make your building tight* and then you assume that machines will do the

DOI: 10.1201/9781003398714-8

rest of the work. But there's more that we can do, and I wonder if we can ask building facades to do more.

Ulysses Sean Vance: I think the more critical aspect of it as well is this sort of idea of 'no maintenance' and the difference between *no-* and *low-* maintenance. How are the strategies of the envelope itself even a mechanism for understanding when, where, and who maintains it? And more importantly, *how?* I understand very holistically that the majority of things that we put into practice are built beyond the understanding of our audiences. There's just a learning curve that we have to be cognizant of. And as the machines become smarter, we assume that they're going to be able to do the work, but we're already at the point where the understanding of maintaining the robot, which is now the house, is far more advanced than those who occupy it. When we look at low-income communities, the economically deprived and disenfranchised communities, they are at the furthest end of that spectrum from being able to maintain these 'smart' houses that we imagine are part of the affordable housing infrastructure. And so how do we either educate them, *or* do we return to a much simpler narrative on building?

Marcel Harmon: A lot of the decisions are driven by the impact on energy. And if we would actually estimate – *quantitatively* – the impacts of productivity and health, that would change the equation quite a bit when we're evaluating these different strategies. Whether it's trying to get air under floors, air distribution system in a facility, or other types of strategies, there's no way that looking at the energy savings is a payoff. But if we look at the productivity and health savings, it becomes a no-brainer, and you get clients who are much more willing to implement this stuff. **If we do a more holistic evaluation of what those impacts are, we get a better result in the end.**

Sarah Haines: We definitely need a more holistic view. We can't just decrease the ventilation to have 'nice and tight' for energy while simultaneously decreasing indoor air quality, which is usually what we find if you tighten the envelope too much. There are pros and cons.

Z Smith: When we were trying to make better buildings, people sometimes resist and say, "the building needs to breathe".

And the problem with that philosophy is that when the temperature difference is not big, that *breathing*, the thing that drives fresh air into your house, doesn't actually do all that much. And what you'll find is that when the weather is really cold or hot outside, your CO_2 levels are pretty good. And when the weather is at all mild, unless you're dutiful about opening the window, you'll find that CO_2 levels are actually quite high, which is bad for your sleep quality and all kinds of activities. **The problem is that relying on leakage to maintain air quality is an inequitable thing because it doesn't work at all weather conditions.** What you want is something that maintains air quality, no matter the weather, but give people that control so that they don't pay a needless penalty in other conditions.

Liz McCormick: That leads into my next question about building automation and trusting the occupant. Are smart, heavily computerized buildings the future after COVID?

Ulysses Sean Vance: **I love smart buildings. I just don't like them in the context where the assumption is that they can cure all.** I think that there's sort of a romanticism, if you will, in our profession that we can actually acutely address every need while in a very obtuse manner [we] give it all over to the robotics. And I think that type of behavior inhibits our ability to understand what is valuable to a condition, an environment, and an ecology. It's very easy to romanticize technology, but maybe we shouldn't. Maybe we should actually ask there to be a much clearer understanding of the boundaries between the wall, the floor, the ceiling, the roof – the layers that actually constitute our assemblies and the mechanisms that we actually provide.

Z Smith: We're doing multi-family housing, and one of the questions is how to make it more accommodating to different people's ways of living. One extreme says to make every apartment its own little island, because if the next apartment over smokes, you want to minimize the chance that air ends up in your apartment. So, if every apartment is on its own, then it can have its own policy, and there's a great attraction to that. The difficulty is that people from different communities have different models of how mechanical systems work. Often, they'll come from a place with a window unit and would turn it off when the conditions were OK. The building was leaky enough that they could get by with that model because there was always fresh air, even when the temperature wasn't bad. But now just by code, we're putting at least Tyvek on buildings – often drunk – but not very tightly. Even badly applied Tyvek will make a house sufficiently tight so that if they *don't* open the window and *aren't* running a mechanical ventilation system, they won't get enough fresh air. So, we've had buildings that were architect-ed and engineered up the wazoo where some people say, "I just don't like the AC, I don't feel good about using it that much" and they'll just switch it off. What we found was the apartment next door loves the AC, and they'll turn it to 62°. What then happens is [the] first occupant is keeping their windows open, and the moist air is coming in. The occupant next door has it at 62°, so now we actually have a condensation plane *between two apartments*, and I have never conceived a world where that would happen. The kind of centralized 'belt and suspenders' approach says to provide neutral dehumidified air to every apartment, whether they keep their windows closed or open, to be sure that there's some degree of air flushing that's happening all the time. And if they want to turn the temperature down or up they can, but I can sleep in good conscience that I provided that. That strategy costs more because I have to run all that duct work, but it makes the building more resilient to different patterns of occupancy. So, I've switched from the John Straub and Joe Lstiburek received wisdom that every apartment really should be independent, to one that makes sure that no matter what they're doing and what their mental model of how things work, we can be confident that they're getting enough fresh air, that the CO_2 levels aren't bad, and that it's drying the apartment so that it doesn't generate mold. So that's changed, even at my advanced age; that's something that I've only come to in the last kind of couple of years from just watching how people occupy spaces differently.

Sarah Haines: I think definitely for new buildings it makes a lot of sense, but we have so much older infrastructure and homes that are already built. How would we implement that into existing homes and existing buildings? I think that's an interesting question as well, of how would that work and is that possible, or are there other things that we need to be solving first before we're turning everything over to AI technology?

Audience (Ken Lambla):   As all of you were speaking, I was thinking about the long-term excuse for not allowing user-occupant fresh air is the imbalance in the rest of the building. And I'm really talking about office, commercial, and maybe medical buildings. My earliest practice days in Chicago were very focused on solar orientations and some adaptive reuse, and man, when we got the owner to allow user affections on the building, it just ruined the system – we were replacing mechanical systems in 15, 20 years. So where are things now? What kind of air balancing goes on as a requirement for sizing?

Marcel Harmon: **Our mechanical engineers will tell you it can be done. It can be done in any building; it's just a matter of setting it up properly from the beginning and it's an issue of control and management.** We're setting up a system where occupant interface interacts with the mechanical system in a manner that doesn't negatively impact it. And so, it will vary somewhat depending on the system type, the nature of the building operations, and what the occupants are doing, *but it can be done.* I'm just thinking in terms of one example where we set it up. It's a K-12 setting where they have natural ventilation, operable windows, and operable vents that everybody can access and turn on. It's set up in a manner that there are indicators when the outside conditions are appropriate relative to the type of system that's been implemented into the building. **The kids love it.** When it's a red light – don't open the windows, don't open the vents; green light – it's great, go ahead, open all to your heart's content. And the kids help manage that system – they're on top of it and it works. It's a net-zero facility where, in addition to the in-room natural ventilation, there's a solar chimney system within the building that works with the under-floor radiant system. And we achieve the net zero targets *and* the high ventilation rates while still achieving those energy performance criteria. At the same time, it's getting rave reviews from the teachers and students and the people that are managing the facilities. It can be done; there are constraints from a climate standpoint that certainly impact some places, but it's not impossible.

Z Smith: UC Merced has a similar system so that every faculty member can open the window in their office. It's turned the other way, though, where the act of opening the window shuts the VAV [variable air volume] box that serves their office. The idea is they're making this active; they're not actually running a control on the VAV box. They think they're just opening the window, but opening the window says, "You're floating with the outside now, son, and this is the world you're in". When you close it, then the AC can come on. The natural act, which is the act of using your body and opening the window, is the way of signaling it. And that's in a climate that can be quite hot. Merced is 100 degrees in the middle of the day in the summer, but it has a lot of the year when it's good. That gets away from unbalancing the air base system. So that's kind of strategy one; strategy two, Marcel just mentioned. Strategy three is what, I think, has shifted in the last 20 years in the proliferation of VRF [variable refrigerant flow] systems, which is just an ecosystem of heat pumps. It's a bunch of little heat pumps that are each saying, "Oh, I need heat, send me some of this. Oh, I need cooling, send me some of that". And what happens is when you decouple the moving of heat from the provision of fresh air, then you're just providing neutral air and you're not unbalancing the system in terms of its ability to control temperature. The air system is then just making sure there's enough fresh air. And it's true that if they open the window and that system is running, some of that nice fresh air is running out the window, but it's okay, because it's just neutral.

I love the example from the original Salk Institute, which is that the way you turned on the lights was if you closed the wooden shutter that blocked off the light in your office. Again, **this direct association between an action that we as humans understand and the mechanical system behind makes people smart.** And I think Don Norman has a great book, *The Psychology of Everyday Things*. And he says, **what we want are control systems that make us feel smart, not control systems that make us feel dumb like that zero-gravity toilet instruction.**

Ulysses Sean Vance: I think what I really like about what both Marcel and Z are proposing is that **it's culturally appropriate, but also behaviorally informative.** And I think in that capacity, that's probably also what I feel changed about the mind state about architecture and mechanical systems, understanding the fact that there is an association between outside light and indoor light as well. Getting to the point where that's actually part of design thinking, that it's not a centralized system where the behavior is taken away from the inhabitant, but one that is actually reflective of their understanding. To me, that's where it gets exciting because there's a certain inclusiveness in that model. We're going to ask the audience, "What are you capable of comprehending given your situation". If you're in a medical environment and you're seeking ownership of the space you're in and even to a certain extent, *authorship* of the space, does it feel like you're home? What are you willing to take on control of and therefore responsibility? And if we think in that capacity, then **we're redefining the relationship between architecture and mechanical systems through culture, and that's where it's exciting for me.**

Audience (Deborah Ryan):   As a landscape architect and urban designer, you made my heart beat when you started talking about the design of parks, walkable places, bikeable places, and 15-minute cities or 15-minute neighborhoods. And I think it's a constant struggle to get architects interested in the space outside their buildings. And so, I was curious as to how you push back on that or have had success with your colleagues to expand the way of thinking about that?

Ulysses Sean Vance: When I talked about the hospital as park in my presentation, we get a lot of pushback from both the architect and the landscape architect. They say, "Well, you can't do examinations or procedure or treatment outside in that space", and I remind them that in tactile healthcare, they do that all the time. They are obviously outside in those conditions, but that's not actually what's exciting about it. The actual space of excitement is the fact that the majority of persons who are avoiding any type of healthcare are actually excited about meeting a doctor who's randomly sitting on a bench in the middle of a park. And that's the place where I get people responding more quickly to this concept is by making the actual medical practitioner excited about it first and therefore realizing that they have a greater capacity for engagement because they're seeking that – they're looking how to make patient awareness and patient engagement their first priority. There are these things called the triple aims of healthcare and they have a quadruple element to them now. But one of the primary things is making healthcare more accessible. The other question, about making architects excited about this, is to realize that it's not actually taking any design capacity out of the problem. It's actually increasing design capacity in the same way where the constraints of the design problem are actually the place where great ideas come from. And I think that's the capacity, at least for me as an architect; I'm hoping that others climb on board.

Marcel Harmon: The other thing is to get more key stakeholders involved early in the process, particularly involving different community members in the process. We have to expand outside the typical focus of the design construction project where we're only thinking about those involved within the building initially. Getting those people involved will bring those concerns up early on so that they can be addressed.

Z Smith: What if you reimagined the thing between inside and outside as the occupiable space? Living in the south we know what that is – it's called a porch – and we know that the porches are often the most interesting things in buildings, the richest places where everyone wants to hang out. There's a term where two communities come together: it's called an *ecotone*. The interesting thing about ecotones is that they are richer in species diversity than either of the places that they draw species from. So, if you have the forest and you have the water, the marsh is the place of the greatest species diversity and the greatest productivity. So, we asked if a building could be that too. When Liz was at EDR, she led a competition team to design the 'next-generation building façade', and we identified that the problem with facades was this separation between inside and outside. It's a mechanical engineering approach to how to make a building façade, which is actually the most interesting part of the building because it's halfway between inside and outside. We designed a spatialized facade system for a high-rise building in Brooklyn. In the winter [it] could help heat the building; in the summer you could dribble water through and help it cool the building. **It has a mechanical function, but it really has a delight function. And as architects, when we're at our best, we're in the delight business.** And so, this was what people were asking for.

Bill Reed, the grandfather of green architecture, plays this game where he has everyone close their eyes . . . [to the audience] close your eyes, close your eyes. Imagine the place where you are happiest right now. Think of a great memory where you were very happy, and it was the best place you could be. How many of you were inside a building when you thought of that moment? The answer is almost no one. And so, if we're in the business of happiness and delight, then maybe we should use that as a way to think through how we automate buildings. I think that unlocks so much hunger in clients for what they really like to see.

Audience (Brook Muller): David Leatherbarrow wrote *Uncommon Ground*, this beautiful book, and there's this chapter where he describes the three architectural horizons. There's the equipment, which is the furniture and the microphone; the practical, which is the envelope largely; and the environmental. And he studies these mid-century modernist architects where the whole architectural expression was the manipulation of the *practical* to mediate relationships between the *equipmental* and the *environmental*. And then I start thinking about these highly environmentally performance-rated buildings that are actually shutting us off from the very thing that they were intended to safeguard. Could we make buildings cheaper that way significantly?

Z Smith: For us, the question is what happens there? And I think there's a reconsideration of post-COVID space, because there's a lot more space in America than we need. Recently, there were some studies at Harvard where they found that faculty members were only in their seats 12% of the hours of the work week. And that turns out to be true, that nowhere in corporate America can you find that people are sitting in their assigned chairs for more than 30% of the hours of a 40-hour work week. People aren't there; they're doing other things. Then the question is, what if we design spaces that are delivering what you really want only where and when you need it? Sometimes you really do need a private office. What if we gave you a private office, but only when you actually needed it? Or if we gave you an amazing space to work in when the weather's fine, like it is right now? If we do that, the building can get smaller. How small could we make buildings if we made each of the spaces incredibly focused on the one thing you want to do there then?

Ulysses Sean Vance: What do you mean by cheaper? Where do you draw the line for cheaper?

Audience (Brook Muller):   The assumption is that green buildings cost a lot more, but I think they could probably cost a lot less if we just made some really good decisions. And if our mechanical equipment was converted to space, you move dollars from utilities to the landscape and how far can we go with.

Ulysses Sean Vance: The reason why I asked is because there's a lot of parallel conditions to when you try to define success. And very often in our field that success is defined by the cost, making it cheaper. And I'm glad that you gave a further context that it's building costs specifically, because the same argument was made about the accessible toilet. And of course, when you look at the lifecycle of that – the lower amount of fatality, the comfort associated with the person, the lack of a need for assisted care – [it] brings a different type of affordance to the thing. So, it might not be cheaper as a product off the bat when you only order five or six of them across a project that has 60 or 70 units, but it could be made more effective and therefore made cheaper. And so, I think that capacity of parallel thinking is established.

Marcel Harmon: Relative to costs, when you're proposing to build your structures, what defines what those costs are to begin with? We live in a free-market economy with very narrow definitions of 'value' that essentially define what things cost and ignore substantial benefits that we get, whether it's equity or health and wellness or myriad other things that actually add real value to our quality of life that are not accounted for in return-on-investment (ROI) analysis or lifecycle-cost-analysis (LCA), and those drive decisions all of the time. **And it's these larger spheres of interaction that will always be barriers to what we can do at the individual project level. So, we can get creative in terms of how we deal with that. Quantifying health productivity is one way to do that.** Until some of those spheres of interactions and how they operate are changed, we will always have those barriers that we are budding up against. And the free-market economy is a major one, and it's an artificial creation that happened back in the Middle Ages and grew from there, but it is not necessarily reflective of reality. *Homo economicus is not real*; it's an artificial construct and we can't change that. If we have the will and the insight to do that as a species . . . I don't know yet.

Z Smith: But I think the first step in accepting the idea that efficiency is how much of what I *wanted* did I *get* and for how much. The wrong way to do that is with the cost per square foot. A better way is with the cost of the project divided by how many people could I provide a great efficient, productive workplace for. Can we construct 'rent' as dollars per person supported as opposed to dollars per square foot? There's no reason we couldn't structure it that way. When you rent an apartment, you actually don't rent by dollars per square foot. You might talk about it – it's a two-bedroom apartment, it's a three-bedroom apartment, but then you kind of look at the totality of the place and you say, "I'll pay that much for that". Strangely, we don't do that for commercial projects or university projects and so on. If we talk about utility divided by cost, then that's a better sense of even in homo economicus financial efficiency.

Audience:

I happen to be a mechanical engineer in the south in a warm-humid climate, and each presentation has been making me more and more uncomfortable; to the point where during your presentation, Z, I started freaking out when you started talking about 20 CFM. To me, that's a lot of air in itself. And then 40 CFM . . . *what the hell is this guy talking about*? And Marcel and Z, I think y'all calmed me down a little bit in your answers when you talk about operable windows, having things in place and when you can't open the windows, when your air conditioning is running and turning it off. And I think what you're alluding to in these buildings where you're getting more outdoor area, you are considering [a] dedicated outdoor air system; it's not a conventional air conditioning system. So, I won't have every architect in Charlotte asking us to design buildings like this. So, I feel I can sleep a little bit better now, except for the first one with the microbes, that still got me. I want to go home and clean the house. We do specify products that have antimicrobial coatings. Have you done any research to know whether those are really effective and if they are, for how long are they good?

Sarah Haines:

That's something I'm currently looking into and we're going to do more research into, because there are so many materials that are marketed as antimicrobial. I think you're all thinking that could be great – maybe it is preventing certain microbes, it's often species-specific or only prevents certain mold growth. When I talked about how everyone's homes are so diverse, you can't just be looking at certain species. But that is to say that if you're adding more chemicals onto a material, that could also be detrimental because you could be creating more exposures. So just because its antimicrobial doesn't mean that it isn't also emitting a high amount of VOCs and ozone. So, in my personal research I haven't, but we're looking into that. I'm working with someone to create antimicrobial coatings for filters (face mask or ventilation), but it's still early stages. I want to make sure that if we are creating some kind of new material or introducing it in a home, that it's not going to be more detrimental. **It's a constant risk calculation of pros and cons of how is this going to be benefit in one way and then harmful in another.**

Z Smith:

I was going to ask in this thing about antimicrobials. What I've noticed in the health industrial complex is vendors are all out there trying to sell us stuff, and you can't blame them, because everyone likes to see a hero when they look in the bathroom mirror in the morning. They all look at what they're selling, whether it's coved flooring or chair rails that are made of weird coppery things that are antimicrobial, they all want to see a hero. I don't blame them; I don't think they're dastardly evil people. I think that it's easy to see something and say to yourself, "Yeah, this is really good. This is great". There's this whole discipline called evidence-based design. What that means is that we have a lot of information about what *actually* impacts the spread of diseases. But from the studies I've read, it's nothing about any of the stuff that the people want to sell us. It turns out to be the curtain that they draw across around your bed is the main surface spreading the microbes. As it turns out, that's the main vector for disease spreading as well as the ventilation system of the hospital. And yet we spend huge amounts of money on all these antimicrobial surfaces all over the building that have relatively little impact. My great hope is that we can learn from the people who are doing evidence-based design in healthcare, and then translate that to what we're doing for our houses and our workplaces.

Sarah Haines:

You're throwing all this money into the surfaces and maybe sterilizing them and then potentially promoting more harmful bacteria or viruses that can then survive in these really niche climates without actually having much impact. Microbes seem to always find a way to survive – they're very resilient.

Audience: I'm tying in with what you're saying and also back to materials, but I'm wondering if at some point we actually figure out DALYs [disability-adjusted life years]. In public health, when you have a bad year of life, it's called a DALY, so you lose quality of life and it's actually quantifiable and materially oriented. So, you can go to the World Bank and say, "I'm going to save so many DALYs per person". It's quantifiable. Can that apply to our interior environment? If we could quantify that if you use some kind of a certain product or lack of a certain products with chemicals that with air quality and so on, you could actually come up with those DALYs, which would lead to grants, monies, awareness by clients – I think it would change it.

Sarah Haines: That is a great point. We talk about this a lot in indoor air quality about how there really aren't many indoor air standards related to exposures. There are a lot of standards that Z was talking about, where you can have *this* much ventilation and *this* is the temperature and humidity, but we don't regulate the level of micro-organisms, VOCs, or other contaminants allowed in buildings. Hopefully this is the way forward to actually create these standards, but it is really difficult to get those levels because buildings are so different.

Ulysses Sean Vance: But I think it's also very important to be aware that some of the problems associated with quality of life are culturally-specific. They are framed by the way in which we associate with end of life as palliative and only a time period associated with the last few years. And to some cultures, those things have a little bit more flexibility and adaptability. For example, in Indian culture, there is a place that you can go when you believe you're going to die. You can hang out there so that you're not in the household, and if it doesn't happen, you simply go home. I have a client in China that is actually working on a similar type of environment that acts more like a spa resort than it does end-of-life palliative care with beeping sounds, machines, and dialysis equipment is all around you. Instead, there's a nurturing, cared-for, enjoyable experience and sort of 'anti-narrative approach' to it. I say that because before we turn it into a material question, it starts to take on what people *perceive* as much as what it is actually *tangible* content.

Marcel Harmon: That's an excellent point about quality of life. What is an acceptable quality of life that is highly culturally variable? We typically talk about the Western concept of quality of life relative to your point. Most of the work that I've done relative to quantifying health focuses typically on things like sick days, insurance claims, or things that are more relevant at the building level. But the idea of trying to actually look at that relative to DALYs is interesting in terms of life because it gets at a quality-of-life issue that some of those other measures don't.

Sarah Haines: With the DALYs, you can do human health risk assessments, which is where those numbers are coming from. You're getting an average daily dose of, say, lead exposure, or there's a whole list that [the] EPA has. Some are cancers, some are noncarcinogens, and there's a different calculation for each. You can also do ecological risk assessment. Perhaps what you're alluding to is if we're using both of those types of assessments, should we be bringing those indoors as well? Maybe we need to be thinking about other exposures that aren't listed by the EPA.

Ulysses Sean Vance: If people are interested in that topic, EDAC [Evidence-Based Design Accreditation and Certification] has a whole section on that. It's a great place to look for those type of metrics and how they're applied.

Audience: Through all of your presentations, there was a common thread of environmental justice as well as the external factors that impact your health and your interior health. To your point about keeping your windows shut, if you live in a community with really bad air quality or bad water quality, or even having a project where all the operable windows are value engineered out, you're already at a disadvantage. All of these are realities that we contend with but have a lot less control over. One question for you, Sarah, specifically in your studies with dust inside of the home being able to tell you about your biome, is there a way to understand if the problems are coming from the exterior factors versus interior building materials?

Sarah Haines: Yeah, absolutely. A lot of other different researchers have been looking into that and using dust as a metric to actually compare different communities and different types of environments to see if it is coming from outdoor systems. Or is it something that's indoor? Most often, we can tell if a microorganism is common outside and if it came inside. Or if you're thinking about chemical volatiles, this is something that definitely is coming from outdoor air. And it seems to be that lower socioeconomic communities have lower air quality just due to systemic problems throughout our society. It's something that I'm currently working on, trying to have equity within our houses and making sure that is something that we're thinking about when going into the communities and talking with people.

Audience: It's easy to feel like it's something we have no control over, and it's very depressing, but I would love to have metrics that we could utilize because **all of this stuff we're discussing today is so invisible and intangible and people often don't know what they're facing.**

Liz McCormick: I want to ask one final question. We talk about the 'return to normal' with COVID, and I think COVID brought up a lot of issues that we already knew we had, especially in relation to buildings and fresh air but also access and equity. Was COVID long enough to really impact us? Did we learn our lesson from the past two years?

Z Smith: I'll start by saying there are a few people who would dearly love to not learn any of the lessons of COVID and just want to 'Sherman, set the way back machine' to March 2020 and we're still going to fight that. I think there are these experiments that have been done that we can't unlearn, as much as people will try to have us unlearn them. But at the same time, I think it's undeniable.

Ulysses Sean Vance: I'm trying not to be the pessimist that I am . . . I really am concerned about this because I think I can look at timelines much broader than the immediate condition and see where our own reflections that led to change were squandered at the expense of sensationalism and popularity. And the most prevalent example is that we had a wonderful time enjoying a transition between power being centralized and power being dispersed in the '60s. And I would argue that the current trend of the Great Resignation is really just a weed-less Woodstock. And for me, the challenge of that is being careful and cautious that the media not be the mechanism to control our thinking as much as we allow it to be. And why I say that is because there was a time period where the power structure learned from the sensationalism that came out of the '60s and learned how to use media to its advantage. We get the '70s and '80s, we get the popularity of content in the '80s. It drives industry to do what it did in the '80s and '90s, to the point where we blame other cultures and countries for our own automation. And the industry itself causes that gap, then that allows us to then have another Great Resignation, which is circular. And it's given to a new generation as if, look it's yours, and in reality it's just another waving of the hand. And I say that because I think that there's great concern for me, at least working in the space of health equity, that what we understood to be the essential worker in their value will return to the dismissiveness prior. That far too easily, we've already started not giving tips; we certainly don't cheer for nurses every day like we did; we're starting to curse them out. But I think even the way in which those we thought were popular cultures seek sensationalism in order to stay alive in this construct where we're all driven by what we see, what we click on; we have to be cautious. And so, I'm not trying to end with the pessimistic note because I think that there's a lot of optimism that is coming out of these movements and it is happening in a circular manner, as long as we keep moving it forward. And that we don't lose a sense of what balance has for us all.

Marcel Harmon: I would share your pessimistic optimism there; I feel the same way. You can just see what's happened here over the last month. People are throwing away their masks; they're getting back together; it's almost like the pandemic is over like it was three or four times before now. And just putting blinders back up relative not just to the potential for the next wave or the next variant, but the conditions that exist to drive the suffering that's occurred throughout the pandemic. Maintaining the status quo and their existing power differentials that are driven by a lot of things, from what I've already mentioned in terms of the free-market economy to existing power structures in place. And I think there has been a hope that some of the upheaval that [has] occurred out of the pandemic may drive some change. I'm still hopeful for that. I don't think I'm as hopeful as I was six months ago, but I'm still hopeful that we can create some of these changes in these higher-level series of interaction that constrain what happens on the day-to-day basis. But if we don't make some changes in that, I honestly don't think there will be significant changes for the better.

Sarah Haines: I agree with what you're saying. If we're thinking more about building science, I would say in a more optimistic sense, indoor air quality is now something that people talk about, which almost no one did three years ago. Now people are thinking about it. So, I'm hopeful that now that it is something that we're taking account of, the conversation will continue. It's just making sure that we are still valuing it and it doesn't just fall off once things return to normal and everyone's like, "Oh, well, now we don't care". So, I am hopeful that things have been changing on that front. And people are interested in having healthy indoor environments.

Liz McCormick: That's a great way to end. Everyone always says, "What do we do next"? And the most important thing to do is start the conversation, which I think is what these four panelists have been doing, and what we're doing here today. We have mechanical engineers here; we have practicing architects as well as students. So, I think this is a really great group, and I'd love to continue the conversation over happy hour. Thank you to our panelists, and thank you all for coming.

8   Reconciling Inside and Out

Elizabeth L. McCormick

Weather is a unifying force. Regardless of our differences, every individual on the planet shares the common experience of weather. Because of its inherent unpredictability, the weather is a continuous source of fascination as well as an accessible topic of casual conversation that is simultaneously polite and impersonal, speaking without saying anything at all. Ever changing, the weather provides a continuous source of stimuli, reminding us of our intimate connection to the natural world. As such, ecology, the study of relationships between living organisms and their physical environment, is inherently a study of the innate heterogeneity of our planet [1]. However, we have engineered the experience of environmental caprice out of the built environment, an action that has actually been regarded as a triumph of modern mechanical conditioning, celebrating man's victory over nature [2]. According to architectural historian and theorist Charles Rice, "being 'exposed to the elements' has become an experience of the inside . . . The weather becomes the defining condition of enclosure, of being inside" [3]. The notion of artificial environments and "man-made weather" has become a hallmark of the air-conditioning era.

Sealed indoor environments, and the homogeneous ecosystems they create, are an entirely human construction. Variability in temperature, light, and humidity all contribute significantly to perceptions of space. However, modern mechanical conditioning strives to produce consistent, uniform conditions over time to achieve a sense of thermal and spatial neutrality [4]. These homogenous environments are purposefully void of stimuli, as Z Smith says, "in pursuit of the *least-offensive* indoor environment" (Smith, p 75). This phenomenon is the architectural equivalent of a spacesuit, another marvel of modern engineering, which is used to maintain homeostasis and protect the human body from environmental extremes [5]. Except in this case, enclosure is used to protect both humans and machines from the unpredictability of weather, as well as the uncertainties associated with outdoor air, which are both thermal and cultural.

Weatherlessness, as discussed in Chapter 2, is both a physical and social phenomenon. The term comes from John Barth's 1958 novel, "The End of the Road", which tells the fictional story of Jacob Horner, a cranky and detached character

DOI: 10.1201/9781003398714-9

who struggles with the concepts of individuality and personal identity. Early in the text, the disoriented protagonist describes his mood as *weatherless*, a "moodless mood, to be without climate, without feeling, freedom or purpose" [6]. This comes from a dream that he had where the chief meteorologist states that there wasn't going to be any weather the following day. "All our instruments agree. No weather" (36). In Barth's world, weatherlessness is not just a thermal condition, but a state of mind and product of a social milieu [7], i.e. the 'empty state of mind' associated with modernism [8]. Ultimately, the goal of this book is not just to support buildings that keep people from being unhealthy, but to provoke a new paradigm of architecture and critical thought that embraces the complexities of the human experience and supports both physical and psychological development.

Binaries and Boundaries

The built environment is embedded in a complex network of binaries: inside versus out, city versus wilderness, and human versus nature, to name a few. While the simple act of identification may offer a sense of clarity and order in an otherwise chaotic world, it can also create divisions between disparate ideas and stifle the richness and vibrancy that emerges from heterogeneity, be it physical, social, or theoretical. Infrastructure systems such as dams, highways, and utility easements, for example, demonstrate how human intervention can divide and disrupt natural flows within ecological networks, constraining interactions between diverse species and ecosystems. Though this book specifically addresses the stark separation between inside and out, and the health and wellness implications it creates, this concept is critical in understanding the larger context of innate human behaviors. The prevalence of these boundary conditions in the built environment underscores the need for a deeper examination of their implications and prompts us to explore alternative approaches that foster connectivity, integration, and inclusivity in architectural design, reflecting the multiplicity of human perspectives and ecological realities.

> **We do not need totality in order to work well.**
> —Donna Haraway, *Cyborg Manifesto*, 1985 (52)

Haraway's statement prompts us to question the necessity of these rigid boundaries. Though "Cyborg Manifesto" was written nearly four decades ago through a post-gender, feminist lens, Haraway's critique of human duality remains relevant in understanding our innate tendency to identify and delineate boundaries and binaries. She describes a cybernetic hybrid of machine and organism, blurring the edges between social reality and science fiction. Her intention is to provoke readers to disassociate from the rigid binaries grounding Western modes of thinking, particularly in regard to humans and nature, as well as humans and machines [9]. Similarly, Félix Guattari describes theoretical concepts that disrupt the notion of singularity and ecological imbalance, highlighting the synergies between human and natural conditions. In his text *Three Ecologies*, originally published in French in 1989, Guattari uses ecological terms to describe humanistic bias, linking three spheres of ecological systems: the environment, social context, and human subjectivity into an ethico-political discourse called *ecosophy* [10] to create a comprehensive and integrated understanding of societal issues. By questioning the ways that entities can blend across different categories, Guattari challenges the traditional understandings of boundary conditions to emphasize the interconnectedness of the various elements of the planet [10,11].

Guattari, Haraway, and others suggest that a progressive society must break down the binaries informing Western ideology, or "Westernity" [12], and exploit the synergies between human and natural conditions, among other cooperative and competing relationships. By embracing the unpredictability of the natural world, it is possible that the inherent heterogeneity of the planet becomes ground for inspiration, not a source of conflict. Perhaps by embracing a more nuanced understanding of boundary conditions, we can cultivate spaces that transcend dichotomies and explore more fluid boundaries that encourage interaction, adaptability, and ecological collaboration where the complexities of human experience can thrive.

In response, some researchers are pushing for a dramatic shift in architectural practice, challenging the very notion of separation, both for energy and health purposes. In the built environment, Haraway's cyborg is Michael Hensel's 'non-discrete' or 'heterogeneous architecture', which describes an environment that is "not merely occupied but inhabited, on not one but many levels, alternately enclosed and open, containing settings and connecting to places beyond the building's assumed limits" [13]. Together with collaborators Jeffrey Turko and Achim Menges, Hensel produced a provocative body of literature in the first decade of the 21st century that explored spatial heterogeneity in architecture with semi-interior conditions that seek to blur the edges between inside and out. Hensel and Menges' 'morpho-ecology' approach, for example, attempts to create synergies between the dynamic conditions of subject, object, and environment [14]. In their 2015 text, *Ground and Envelopes: Reshaping Architecture and the Built Environment*, Hensel and Turko explore the spatial elements of perceived boundary conditions in architecture in direct opposition to the modern fixation with *objects* [13]. Together with Christopher Hight, Hensel and Menges edited a compilation of key texts in "Space Reader: Heterogeneous Space in Architecture" (2009), which includes some of the most relevant voices of the air-conditioning era, including Reyner Banham, Peter Sloterdijk, Robin Evans, Gilles Deleuze, and Félix Guattari. The contributors offer critical insights into the evolving nature of space, exploring the potential of architecture to create engaging, interactive, and inclusive spaces. By compiling this group of essays in a single space, it offers a rich compilation of ideas, theories, and case studies meant to inspire architects, designers, and researchers to reimagine spatial design and embrace the dynamic and diverse possibilities of heterogenous space.

Case Study: A Heterogenous Approach

This section describes EskewDumezRipple's (EDR) award-winning entry in the 2018 *Metals in Construction* Magazine's Next Generation Facade Design Competition, based on the philosophy that the 21st-century facade needs to be responsive to human needs, especially in the workplace.[i] The competition brief called for a 30-story office building located in Brooklyn, New York, directly outside the rapidly evolving Brooklyn Navy Yard and adjacent to the East River.

The goal of this project was to critically question the role of the façade as *separator*, as discussed throughout much of this text and shown in Figure 8.1. Some high-performance buildings have employed a double-skin façade, which represents the techno-centric approach to sustainable design. Though this strategy may help to reduce energy consumption in certain climates, it still provides a homogenous interior space. Instead, EDR's proposal, entitled *The Ecotone*, offered a different approach to the façade as boundary condition and created a diverse community of occupiable microclimates that blur the boundaries between inside and out.

Ec·o·tone (noun): a region of transition between two ecosystems

An ecological niche describes a distinct set of environmental circumstances where certain species often emerge by adapting to the conditions of the niche.

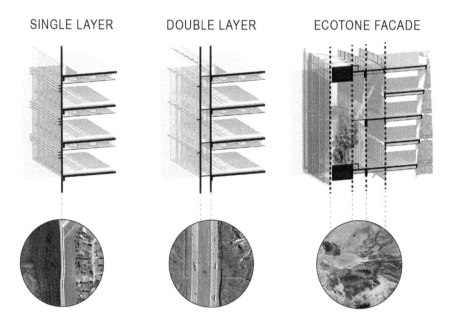

SINGLE LAYER   DOUBLE LAYER   ECOTONE FACADE

Figure 8.1 Conceptual diagram showing the difference between traditional façade strategies and an ecotone facade. Image courtesy of EskewDumezRipple.

These conditions, however, are not static, and depend on many factors beyond each individual species. Where niches overlap, competition occurs, and it is at this in-between (mixing) zone where opportunities for ecological richness and productivity are most potent. Instead of a thin line between inside and out, *The Ecotone* proposed a diverse community of layered thermal and spatial gradients that blend the exterior and interior environments.

ASHRAE identifies New York City as a 'mixed' climate, meaning it faces extremes of both hot/cold and humid/dry conditions. The facade had to be adaptable in order to respond appropriately to both changing outdoor conditions and interior programmatic desires. In summer conditions, shown in Figure 8.2, the layers are largely peeled open and south-facing exposures are shielded from solar heat gain by deciduous plants along the edge of the building. Outside air is evaporatively cooled as it passes through a misting mesh and a layer of open pivoting windows into the planted zone. It then passes freely through openings in sliding partitions into a zone where water cooled below the dewpoint of the air is to dehumidify the air as water vapor from the air condenses on the cooled water droplets. Finally, the pre-conditioned air passes through a series of water source heat pumps above the innermost layer of glazed pivot doors, where it is cooled to its final temperature and delivered via floor registers to the level above. The summer winds from the south maintain positive pressure to keep the outside air moving through the system. Warm stale air is drawn up into the

Figure 8.2 Summer Condition. Image courtesy of EskewDumezRipple.

large rooftop greenhouse through exhaust tubes, which also contribute to the structural system of the building. The rooftop greenhouse allows this exhaust air, which vents naturally upward due to the stack effect, to dissipate naturally into the atmosphere. This lush rooftop space, shown in Figure 8.4, also serves as an amenity to building tenants to take a break in a landscape not typically offered in urban environments.

During the winter, shown in Figure 8.3, the perimeter layer of pivoting glass and sliding panels separating the planted zone from the water feature are closed, transforming the planted zone into a passively heated, insulating solarium. A small amount of outside air is allowed to flow behind the pivoting glass panels to maintain fresh-air circulation in this zone. This zone serves to buffer between various interior conditions and potentially harsh winter weather. The greenhouse's operable roof panels are closed and the exhaust air recirculates in the greenhouse, where the vegetation oxygenates and cleans the recycled air, using plants to serve the role traditionally assigned to machines. Additional fresh air is added and pre-conditioned as necessary through an enthalpy recovery ventilator (ERV). The now-clean and warm air from the greenhouse is then recirculated into the waterfall zones on each floor, which are now used to humidify the dry winter air. It then moves through heat pumps back into the interior zone similar to the summer condition. To account for any additional heating demand, low-energy radiant floors provide the remaining heat.

Figure 8.3 Winter Condition. Image courtesy of EskewDumezRipple.

Figure 8.4 Rooftop Greenhouse. Image courtesy of EskewDumezRipple.

In short, the façade acts as a thermal *gradient* during summer months and becomes a spatial *buffer* in the winter, using transitional space to temper extreme conditions and connect humans to outdoor air. Within this approach, a system is created that minimizes the abruptness of change in environment across the series of layers, reducing the energy required to maintain the system. The building becomes an armature supporting human activity in its most natural context, systemically integrated with its surrounding environment. Acoustic comfort is also addressed by planted buffers and waterfalls, and visual access to the East River and city are a crucial part of the formal response of the project. Users can move through a variety of microclimates and modify their environment on each floor, as shown in the thermal gradients of Figure 8.5. A key consideration in the development of *The Ecotone* involved the design of spaces for many different environments and, by proxy, different individuals. The Ecotone and its spaces evolve and change throughout the seasons. It is architecture by way of transformation. And via this transformation, the structure affords residents and visitors the opportunity to find their own space within. While appropriate for the emerging mobile style of working, this diversity is also imperative for occupant well-being.

As competition juror Aulikki Sonntag stated:

Despite the technical holes in the project, the key thing is: thermodynamics didn't change. And they will not change. Hot air rises, there's a benefit of thermal mass . . . the building relies on the basic laws of thermodynamics. *The spirit is there.* [15]

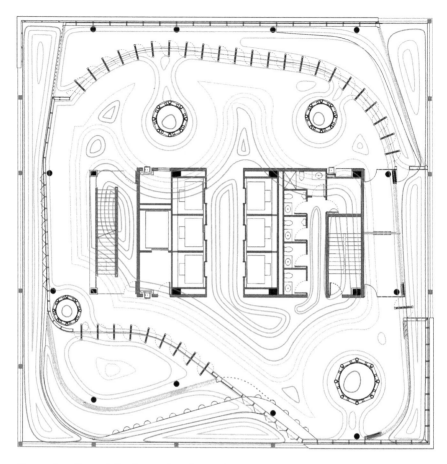

Figure 8.5 Thermal Gradients of Typical Office Layout. Image courtesy of EskewDumez Ripple.

While this strategy relies on a technological approach, it does so to challenge the modern paradigm that sustainable, healthy buildings are only those that isolate the occupant from the outdoors. It does not suggest a return to natural conditions or vernacular architecture. Instead, it proposes a deeper connection to the natural environment through design, balancing advanced mechanical technologies with thoughtful spatial articulation. That being said, this project was a short, theoretical exercise conducted entirely within the confines of an architectural design practice and would require significant collaboration with many different disciplines to develop and formalize the strategy.

Toward a Heterogeneous Design Methodology

> Architecture is too important to leave solely to architects . . .
> —Ole Bouman, architectural historian [16]

In the broader view of global sustainability and human health, architecture exists in an interdependent web of cause-and-effect conditions that stretch well beyond its immediate disciplinary boundaries. The first chapter of this book opened with a quote from Dr. Margaret Chan, Director General of the World Health Organization, who said that the health concerns associated with air quality cannot be addressed using "conventional tools", but that all industries need to get engaged [17]. At a larger scale, the notion of homogeneity in the built environment addressed in this chapter extends to a common phenomenon observed in professional discourse – disciplinary silos – which are characterized by perceived boundaries between responsibilities, objectives, and liabilities. This professional fracturing restricts the exchange of ideas between disciplines and limits holistic problem-solving approaches. Instead, breaking down these barriers and fostering interdisciplinary cooperation is crucial for addressing complex challenges and fostering a more interconnected and dynamic intellectual landscape. In order to holistically address health and wellness in the built environment, designers must recalibrate the way that we think, frame questions, and collaborate with others. A non-disciplinary systems approach to sustainability and human health can better address the fundamental needs and integrity of a system in context.

It is important to develop a comprehensive understanding not only of ecological systems, but of the cultural context in which these networks are supported, questioning the way that humans *exist* in the natural world, not just the way we *manipulate* it. Many of the tools that we still rely on today, including electricity, automobiles, and air conditioning, were developed as systems to dominate nature. By recalibrating the way that various disciplines interact, it may be possible to revisit practices and paradigms in both the social and natural sciences, responding to both environmental and social demands concurrently. In fact, it is at this professional mixing zones, similar to the ecological condition referenced in *The Ecotone*, that the opportunities for innovation in design are greatest.

Sustainable systems are complex, with many moving parts that continuously interact and self-organize. Compounding the issue, researchers are often working with *inherited* problems, research questions, and methodologies. In both industry and academia, for example, there is a growing emphasis on high-tech solutions and building optimization. While this approach helps to reduce the intensity of the impact, **it does not question the way that humans occupy the planet, nor does it challenge our authoritarian, one-sided relationship with nature.** Additionally, long-term energy efficiency practices do not historically lead to overall reductions in consumption. Instead, technological improvements in efficiency typically lead to *increases* in consumption, rather than decreased usage [18]. Known as Jevon's Paradox, this phenomenon has occurred with countless technological advancements throughout history, most recently including renewable energies, energy-efficient lightbulbs, and electric vehicles, none of which actually question the role that these technologies play within buildings and cities. Beginning with the control of fire, humans have manipulated nature to address anthropocentric needs, using technology where nature alone would not suffice.

Ultimately, the techno-centric approach to sustainable development as the dominant model of innovation in industrialized countries has privileged the distributions of energy, knowledge, and opportunity, preserving existing power structures of political and professional elites. As buildings become more technologically complex, performance decisions are often relegated to the opaque domain of engineers and scientists, further perpetuating disciplinary silos and sociotechnical divides. Because technology plays such a major role in modern society, the public is forced to rely on professionals with specialized skills to innovate, design, manufacture, and manage increasingly complex networks [19]. Instead, the infrastructure that corporations have established to promote professional efficiency is the exact organizational scaffolding that constrains disruptive change. Forty years ago, Donald Schön writes that Professionals seemed to have a vested interest in prolonging the conflict. A series of announced national crises – the deteriorating cities, poverty, the pollution of the environment, the shortage of energy – seemed to have roots in the very practices of science, technology, and public policy that were being called upon to alleviate them. Little has changed. [20]

In this text, we do not speak of the potential of buildings in a nostalgic sense – we do not hope to return to a pre-Industrial Era or vernacular architecture. We *embrace* technology. We take a historically informed, progressive approach that acknowledges the historical influences on our perceptions and understanding of human space, but embrace the potential of designing *with* technology, not *for* technology. As the climate spirals further into crisis caused by human behavior, architects and designers must disrupt the existing norms of professional practice and academic discourse, and even public policy, to put human occupation in better alignment with the natural rhythms of the planet.

Ultimately, this book straddles many binaries to explore traditions that blur edges and go beyond the bipolarity of Western ideologies through a recalibration of social, spatial, and environmental behavior. The goal of this text is to promote design changes that purposefully strengthen the human relationship with nature while also destigmatizing the non-human elements of architecture. Neither human nor environmental challenges can be solved by technology or behavioral change alone. By understanding the relationships between the built environment, technology, and human behavior, our hope is to provoke a new perspective of heterogeneous, adaptive, and layered architecture. We must transcend this dichotomy and harness the synergies between technical and social change.

Note

i The author was a member of the design team for this competition during a research fellowship at EskewDumezRipple. In addition, the EDR project team included Tom Gibbons, Mike Johnson, Logan Notestine, Christian Rodriguez, and Mark Thorburn, with special support from Noah Marble, Javier Marcano, Sam Levison, and Z Smith.

References

[1] Hight, Christopher, Michael Hensel, and Achim Menges. 2009. "En Route: Towards a Discourse on Heterogeneous Space beyond Modernist Space-Time and Post-Modernist Social Geography." In *Space Reader: Heterogeneous Space in Architecture*, edited by Christopher Hight, Michael Hensel, and Achim Menges. AD Reader. Chichester, UK: Wiley.

[2] Cooper, Gail. 2002. *Air-Conditioning America: Engineers and the Controlled Environment, 1900–1960.* New York: JHU Press.

[3] Rice, Charles. 2009. "The Inside of Space: Some Issues Concerning Heterogeneity, the Interior and the Weather." In *Space Reader: Heterogeneous Space in Architecture*, edited by Michael Hensel, Christopher Hight, and Achim Menges. AD Reader. Chichester, UK: Wiley.

[4] Luo, Maohui, Richard de Dear, Wenjie Ji, Cao Bin, Borong Lin, Qin Ouyang, and Yingxin Zhu. 2016. "The Dynamics of Thermal Comfort Expectations: The Problem, Challenge and Implication." *Building and Environment* 95 (January): 322–9. https://doi.org/10.1016/j.buildenv.2015.07.015.

[5] Hight, Christopher. 2009. "Putting Out the Fire with Gasoline: Parables of Entropy and Homeostasis from the Second Machine Age to the Information Age." In *Space Reader: Heterogeneous Space in Architecture*, edited by Michael Hensel, Achim Menges, and Christopher Hight. AD Reader. Chichester, UK: Wiley.

[6] Barth, John. 1967. *The End of the Road.* Rev. ed. Universal Library, UL 240. New York: Grosset & Dunlap.

[7] Stuhr, John J. 2019. "Weatherlessness: Affect, Mood, Temperament, the Death of the Will, and Politics." *DisClosure* 28 (January): 1–11. https://doi.org/10.13023/disclosure.28.06.

[8] Pope, Albert. 2009. "Mass Absence." In *Space Reader: Heterogeneous Space in Architecture*, edited by Michael Hensel and Christopher Hight. AD Reader. Chichester, UK: Wiley.

[9] Haraway, Donna Jeanne. 2016. *Manifestly Haraway.* Posthumanities 37. Minneapolis: University of Minnesota Press.

[10] Guattari, Félix. 2005. *The Three Ecologies.* London; New York: Continuum.

[11] McGaw, Janet. 2017. "Machinic Architectural Ecologies: An Uncertain Ground." In *Architecture and Feminisms.* London: Routledge.

[12] Kipnis, Jeffrey. 2009. "Towards a New Architecture." In *Space Reader: Heterogeneous Space in Architecture*, edited by Michael Hensel, Christopher Hight, and Achim Menges. AD Reader. Chichester, UK: Wiley.

[13] Hensel, Michael, and Jeffrey P. Turko. 2015. *Grounds and Envelopes: Reshaping Architecture and the Built Environment.* London: Routledge, Taylor & Francis Group.

[14] Hensel, Michael, and Achim Menges. 2008. "Designing Morpho-Ecologies: Versatility and Vicissitude of Heterogeneous Space." *Architectural Design* 78 (2): 102–11. https://doi.org/10.1002/ad.648.

[15] Pawlynsky, Areta, Melissa Wong, Illana Judah, Yalin Uluaydin, Tali Mejicovsky, Aulikki Sonntag, and Phillip Anazalone. 2018. *Public Jury Discussion for the Metals in Construction Magazine, Design the Next-Generation Facade Competition.* New York, NY: TimesCenter.

[16] Harrouk, Christele. 2021. "'As Long as There Are Human Beings and Their Challenges, There Will Be Architecture': In Conversation with Ole Bouman." *ArchDaily*, April 9. www.archdaily.com/959771/as-long-as-there-are-human-beings-and-their-challenges-there-will-be-architecture-in-conversation-with-ole-bouman.

[17] Climate & Clean Air Coalition, dir. 2016. "Dr Margaret Chan, WHO Director General, Message to the UN Environment Assembly." www.youtube.com/watch?v=tQwXjvQsQ4I.

[18] Jevons, William Stanley, and Alfred William Flux. 1865. *The Coal Question; an Inquiry Concerning the Progress of the Nation, and the Probable Exhaustion of Our Coal-Mines.* New York: A. M. Kelley. http://archive.org/details/coalquestionani00jevogoog.

[19] Brand, Ralf, and Andrew Karvonen. 2007. "The Ecosystem of Expertise: Complementary Knowledges for Sustainable Development." *Sustainability: Science, Practice and Policy* 3 (1): 21–31. https://doi.org/10.1080/15487733.2007.11907989.

[20] Schon, Donald A. 1991. *The Reflective Practitioner: How Professionals Think in Action.* London; New York: Routledge, 2017. https://catalog.lib.ncsu.edu/catalog/NCSU4682382.

Acknowledgments

In the summer of 2021, I had just completed my first year as a part-time doctoral student at the North Carolina State University College of Design and was in the midst of my fourth, maybe fifth, research topic change. Immersed in the second year of the COVID-19 pandemic, I, like others, was becoming acutely aware of the health implications of the indoors as restaurants moved to al fresco dining and classrooms shifted outdoors. With a background as a Passive House consultant, my professional experience had been guided by airtight, energy-efficient design. But as my career progressed, I developed a lingering skepticism about the emphasis on airtightness and mechanical ventilation in sustainable building design, particularly in regard to human health and our relationship with the planet. I was left with the constant wonder of how to balance human health with building performance.

It wasn't until I saw the vintage Cellophane advertisements in Nancy Tomes' "The Gospel of Germs" that I fully recognized the profound influence that social and cultural constructs have on the way we design and condition buildings in the United States. From that pivotal moment, the research question crystalized, and I became consumed with the untold stories of these archival photos. What now exists as an edited book once began as a humble conference paper presented at the Architecture Research Centers Consortium (ARCC) conference in Miami in 2022, co-authored with my now-dissertation advisor, Traci Rose Rider. I am deeply indebted to Traci's support (and patience) throughout this journey.

I am also grateful to the University of North Carolina, Charlotte School of Architecture for supporting the Inside/OUT symposium in the spring of 2022 and Director Blaine Brownell's encouragement in compiling the outcomes into a publishable work. The symposium brought together a diverse group of experts of the indoors, fostered discussions, generated ideas, and fueled the creation of this book. To Sarah, Marcel, Z, and Sean, the contributors who dedicated their valuable time and energy to this endeavor, I extend my heartfelt appreciation. Your work inspires and offers much-needed visions for a post-pandemic world, where health, wellness, and equity are at the forefront of our design decisions.

The evolution of this project serves as a testament to the diverse range of influences that shape our built environment and the inherent connections between architecture, technology, and culture.

This project was a collaborative effort fueled by the passion and dedication of many individuals. To all those who have contributed to the realization of this work, big or small, I am eternally grateful. I would like to express my sincere gratitude to the UNC Charlotte and NCSU libraries, whose unwavering support facilitated the acquisition of numerous invaluable resources included in this book. The tireless efforts of the librarians and staff enabled me to delve deep into the wealth of knowledge and materials that have enriched the research behind this book. I am fortunate to be a part of the UNC Charlotte faculty as well as the NCSU doctoral program, both of which have provided an invaluable network of support and camaraderie. I am inspired by my colleagues in both venues. Thanks especially to M. Elen Deming for the invaluable editorial guidance throughout the writing process. Her expertise, patience, and mentorship not only helped to polish the text, but also ignited within me a genuine love for the art of writing.

Finally, I would like to express my deepest appreciation to my family, especially my parents Barb and Dan, for their unwavering support and encouragement. Their support has been the foundation for this project, among many others. Thanks also to my siblings Kate, Andy, James, and Lauren for their tireless support and encouragement during the challenging moments along the way.

Index

For Product Safety Concerns and Information please contact our EU
representative GPSR@taylorandfrancis.com
Taylor & Francis Verlag GmbH, Kaufingerstraße 24, 80331 München, Germany

www.ingramcontent.com/pod-product-compliance
Ingram Content Group UK Ltd.
Pitfield, Milton Keynes, MK11 3LW, UK
UKHW020936280425
457818UK00036B/343